Plant Bodies

Jitka Klimešová

Plant Bodies

How Plant Structures Contribute to Our Comprehension of Their Life-History Strategy

Springer

Jitka Klimešová
Institute of Botany
Czech Academy of Sciences
Průhonice, Czech Republic

ISBN 978-3-031-92191-9 ISBN 978-3-031-92192-6 (eBook)
https://doi.org/10.1007/978-3-031-92192-6

Translated from Czech

The original submitted manuscript has been translated into English. The translation was done using artificial intelligence. A subsequent revision was performed by the author(s) to further refine the work and to ensure that the translation is appropriate concerning content and scientific correctness. It may, however, read stylistically different from a conventional translation.

Translation from the Czech language edition: "Těla rostlin. Jak nám tvar rostlin pomáhá v poznání jejich životní strategie" by Jitka Klimešová, © Autorský kolektiv Botanického ústavu AV ČR, v.v.i., 2022. Published by Academia. All Rights Reserved.

This Springer imprint is published by the registered company Springer Nature Switzerland AG
The registered company address is: Gewerbestrasse 11, 6330 Cham, Switzerland

I would like to dedicate this book to my father and first teacher of botany, Vojtěch Piňos.

Foreword

Wrapping your mind around the dizzying diversity of plant architecture and growth forms can be a daunting task. At first glance, it seems like a straightforward exercise of simply categorizing species into distinct groups such as trees, shrubs, and herbs. But what about cacti and aquatic plants, how are they to be classified? Many species seem to break the rules. When you commit to deeply exploring this wonderful world of botany in detail, the number of rule breakers appear to multiply exponentially. Because of this, plant growth form classification schemes blossomed into a cottage industry with little consensus, and the secret to understanding the evolutionary basis of plant architecture remained shrouded in mystery. Solving this mystery is the basis of this book.

The author, Jitka Klimešová, learned right away that the mystery could not be solved by simply looking up taxonomic descriptions of species. Such descriptions were not just incomplete, they were contradictory. Even basic theories for explaining the origin of different plant organs had not been fully resolved. Just when we thought that leaves, stems, and roots were distinct organs, botanists discovered too many mischievous species that blended their organs into continuous features. As is often true in dynamic fields of science, our original ideas did not survive scrutiny. Many botanists around the world have worked hard to propose generalizable frameworks to make sense of the diversity of plant forms. Many of these pioneering scientists are characters in this story and are memorialized artistically by the author.

Students of botany have lacked a standard reference to start their journey toward understanding the architecture of plants. Thankfully, Jitka Klimešová has synthesized this scattered literature into a framework that can be used by botanists around the world. Importantly, she moves the discussion from descriptive observations toward an enterprise to understand how this diversity of morphology drives variation in function. This book marks the beginning of an important field of research she calls functional morphology, with a call to go belowground and integrate what we have learned aboveground into a comprehensive understanding of the whole plant body.

Department of Botany
University of Wyoming,
Laramie, WY, USA
27th April 2025

Daniel C. Laughlin

Preface

When I delved into my self-study of plant morphology, it was with a clear goal to accurately describe the diversity of clonal growth organs among plants of Central Europe. Initially, I imagined that I would simply find descriptions of clonal growth organs for individual species in books and list them in a table. Based on the data from the table, I wanted to calculate the types of morphological categories and the number of species in them. I went through a lot of literature published from the mid-nineteenth century to the present, written in several languages. I started with a Russian-written ecological flora of the Moscow region, then I moved on to English and German literature. The first problem was that species for which I found multiple descriptions often had different descriptions in each source, and the second was that the number of categories was enormous. There are many classifications of morphological categories, and authors apply them more or less carefully to describe species. To cope with these problems, I started collecting plants and tried to understand their morphology on my own. For this purpose, I began drawing plants, especially their belowground organs, which I aimed to understand. Drawing became my passion. I spent every free moment digging up and drawing plants. I was on maternity leave at the time, so in the morning, during walks with the children, I collected plants, and when they napped after lunch, I drew them.

When I returned from maternity leave, I had collected material for our entire flora. In 2003, we were invited with colleagues to participate in building a European database of plant functional traits. Within this project, I collected more data, clarified how to define the traits I wanted to describe on plants, and filled the CLO-PLA database, which is freely accessible on the internet. My interest in plant shapes could have ended there, but I gradually began to realize how little knowledge of plant morphology there is among ecologists and how this lack of knowledge is limiting their ability to understand various problems in plant ecology. This led me to start a course on Plant Architecture. However, only a few students take such a course, and I was not satisfied with that. With colleagues, we established an international week-long course for young scientists, but even that is not enough. I decided to write my course as a book, which I hope will be read not only by students but also by researchers who deal with plant ecology. Hopefully, this book will help them better understand plant shapes, and the dependent ecological functions, and will inspire them in their own research.

Průhonice, Czech Republic Jitka Klimešová

Acknowledgments

The book would never have been created without joint projects and fruitful discussions with many students and colleagues. My thanks go especially to Alena Bartušková, Jiří Doležal, Arinawa Filartiga, Zdeněk Janovský, Vít Latzel, Jana Martínková, Gianluigi Ottaviani, and Curtis Lubbe. For reading previous versions of the manuscript and providing constructive comments and suggestions, I thank Tomáš Herben, Adam Klimeš, Zdeňka Lososová, Silvia Pecháčková, and Curtis Lubbe (reading English version). For the illustrations and photographs, I thank Alena Bartušková.

I am indebted to Zuzana de Ruiter for help with English edition of the book.

The book could not have been created without the support of grant agencies for the work of our research group. I thank the Grant Agency of the Czech Republic (19-13103S, 19-13231S), the INTER-TRANSFER Program, Ministry of Education, Youth and Sports (LTT20003), Praemium Academiae award and the long-term development project of the Academy of Sciences of the Czech Republic [No. RVO 67985939], and the Institute of Botany of the Czech Academy of Sciences.

Competing Interests The author has no competing interests to declare that are relevant to the content of this manuscript.

Introduction

Ecologists have a hard time dealing with plants and must be very inventive in their research. Everything in the plant kingdom takes a long time and is often hidden belowground. With plants, we cannot compare strategies by describing what prey they hunt or how much time they spend on various daily activities; we often help ourselves by attributing functions to plant shapes. Surely you agree that a tree, shrub, or herb have different life strategies, or that a leaf, stem, and root have different functions. But what about different types of stem branching? Does it also have any ecological significance?

I want to introduce you to the shape of plants and its significance for plant ecology here, with an emphasis on seed plants and their vegetative organs (flowers, fruits, and seeds will only rarely be mentioned). In the first part, I will focus on the overall appearance of the plant: the growth form. In the second part, I will focus on comparative morphology, that is, on the basic organs, such as the root, stem, and leaf, and their transformations and functions. In the third part, I will discuss recurring motifs in the structure of the plant body: buds, twigs, and shoots, that is, the architecture of the plant. The aim of all this is to show what we already know about the relationship between form and function, to reveal what we do not know, and to suggest how we might further use information about the shape of the plant body in ecology.

The chosen approach has its advantages and disadvantages. As an advantage, I can present ecological questions from an unusual angle, to show connections that remain hidden if you do not know how the plant body is constructed. The disadvantage of this approach is that some topics—ecological functions such as regeneration after disturbance or clonal growth—will be discussed repeatedly. For example, I will mention the reaction of plants to disturbance in my discussion of growth forms because their definition is partly based on how plants react to disturbance. I will return to this topic in the chapter on morphology, because some belowground plant organs serve for regeneration after plant injury. Finally, I will mention disturbance in the chapter on architecture because the supply of buds from which the plant regenerates is largely determined by its architecture. I believe that these repeated returns to a certain topic will help present the ecological function each time in a new light and will not be too disruptive.

The path to understanding plant shapes and their function relates to the turning points in the lives of researchers, and these are often influenced by historical events. Therefore, I have taken the liberty to supplement the text with brief profiles and illustrations about the lives of some researchers who have studied the shapes of the plant body.

Contents

About the Author

Jitka Klimešová was born in 1963 in Rýmařov, Czech Republic. She is a plant ecologist interested in plant functional morphology. She studied Geobotany at Charles University and did her PhD at the Institute of Botany, Czech Academy of Science where she is employed from 1993. She also lectures at University of South Bohemia in České Budějovice and at Charles University in Prague. She has published two books *Temperate Herbs: An Architectural Analysis* in 2018 and *Těla rostlin. Jak nám tvar rostlin pomáhá v poznání jejich životní strategie* in 2022. Her research group studies belowground plant organs, creates open access global database of plant clonal traits, and organizes practical course for assessing plant belowground traits Go_Belowground! She is a devoted popularizer of science writing blog and contributes to Czech journals popularizing science. Her hobby is graphical art, and she uses her linocuts to illustrate her texts.

1 Growth Forms of Plants

Early ecologists believed that growth form categories are crucial tools for ecological research. For distinction of plant growth forms, the whole plant appearance is used. There are many classifications of growth forms that may have a few or many categories and either describe diversity of plant growth habits per se, or describe categories that are considered as adaptations to some environmental gradient (e.g. climate or soil) and are called life-forms. The most used classification was developed by Christian Raunkiær and defines life forms that represent plant adaptation to survive a seasonal adversity. The life forms are classified according to position (in respect to the soil surface) of the highest bud that serves for resprouting after seasonal rest. According to Raunkiær's original idea, the spectrum of the forms in a community should be compared with the theoretical spectrum of the whole world flora. Currently, rather than comparison with a theoretical spectrum, the spectra found in individual communities are compared with one another. The functionality of life form or growth form categories is only rarely experimentally tested.

1.1 How Do We Perceive the Plant as a Whole?

1.1.1 Basic Growth Forms

Tree, shrub, and herb are the basic growth forms that each of us commonly recognizes. These growth forms simultaneously represent three strategies of plant growth concerning the likelihood of strong destructive disturbance. Herbs do not take the colonization of above-ground space too seriously, producing only soft, short-lived bodies that are inexpensive, thus thriving well where some disturbance regularly occurs. This inexpensive herbaceous body above-ground is destroyed by disturbance, and the plant regenerates from belowground organs or seeds once the disturbance subsides. Shrubs invest more in aboveground biomass, and it lasts longer for them, but they keep some resources stored in belowground organs as a precaution. When their above-ground biomass is destroyed, the belowground organs survive

J. Klimešová, *Plant Bodies*, https://doi.org/10.1007/978-3-031-92192-6_1

and give rise to new above-ground biomass. Trees invest in greater height, do not form belowground storage and regenerative organs, and therefore grow in places where disturbance is very rare; if a strong disturbance occurs, they usually do not survive it. If in a climate where trees can grow, a plant community is composed only of herbs, you can be sure that above-ground disturbance occurs frequently there. This is the case with Central European meadows, which would become overgrown with forest if we did not mow them.

Basic growth forms are also limited in their occurrence by other factors such as climate or soil properties. Herbaceous biomes, like tundra, savanna, or steppe, exist where trees cannot grow due to cold or lack of precipitation, although disturbance by fire and grazing plays an important role in savannas and steppes. Shrub-dominated biomes are usually plagued by a lack of water and nutrients, and the presence of fires. Forest biomes are usually undisturbed or experience catastrophic disturbances only very rarely (Fig. 1.1). Human activity has, of course, significantly impacted the distribution of growth forms, mainly because disturbance is associated with human activity. Due to intensive agricultural use of land for growing short-lived crops, grazing, and urbanization, original forests, steppes, and savannas are disappearing. Conversely, in some steppes and savannas, shrubs and trees are gradually spreading after the exclusion of herbivores and fires. Climate change, in turn, supports the spread of shrubs into the tundra due to warming.

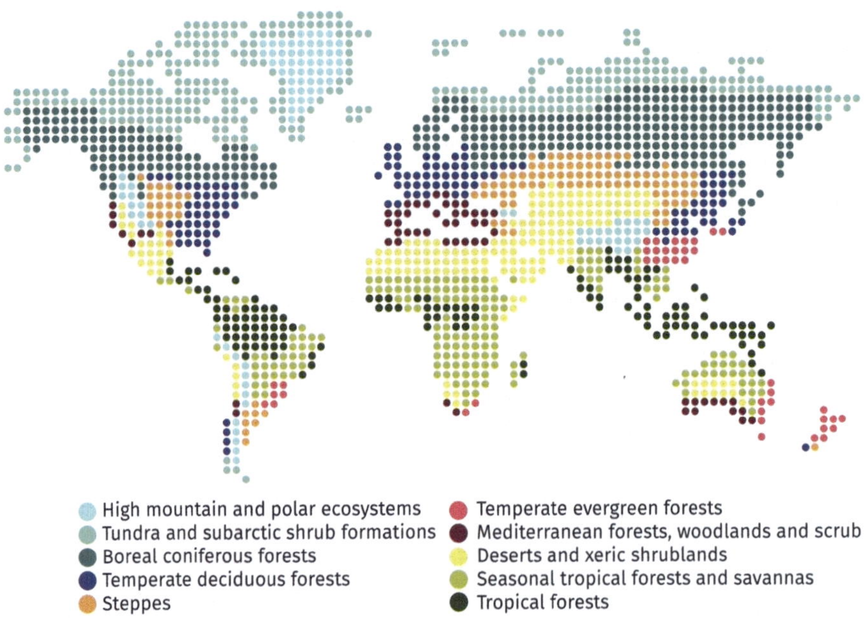

Fig. 1.1 Distribution of the main biomes on Earth. Visiting a point marked on the map as a savanna or deciduous forest does not necessarily mean you will see that community; you will likely be surrounded by arable land and will need to visit a national park to experience that biome

Herbs, shrubs, and trees often occur together in one community. Such a community, for example, the deciduous forest, is not disturbed (hence the presence of trees), but it also provides space for other growth forms, often not only herbs and shrubs. Particularly high diversity of growth forms is possible where the boundary between above-ground and belowground, which plants need for rooting, is not just the soil surface. Many rooting opportunities exist, for example, in tropical rainforests, where small plants can grow on large plants. Additionally, there is sufficient moisture, and the fall of leaves and twigs, which get caught in the forks of large trees, creates favorable conditions for root establishment. Similar rooting opportunities are provided by aquatic environments: some plants root in the substrate at the bottom, while others manage by floating in the water column or on the surface. In both these environments, the tropical forest and the shallow water reservoir, there is great competition for light, and we find growth forms that do not invest in supportive structures, such as slowly growing woody stems, but rely on the support of the water column or parasitize on the structures of surrounding plants. With the help of twining stems, adhesive roots, or thorns, lianas climb over surrounding plants to reach better light conditions.

From the perspective of the temperate zone with a strongly seasonal climate, the division of plants into trees, shrubs, and herbs seems straightforward, but in biomes where seasonality is less pronounced, the defining growth forms is not so simple. Some plant characteristics that inhabitants of a temperate climate associate with herbs, such as small size and soft tissues, may not be sufficient for identifying members of the herbaceous growth form category. Why? Well, perhaps because a tall plant does not necessarily need to have a woody stem (hard and lignified) but instead can have a softer body and just be a large herb (for example, a banana plant). If we define basic growth forms by their size, large tropical herbs will belong to the "tree form"; if we add the condition for the "tree form" to have a woody stem, we will not be able to include these large herbs in the system unless we remove the size condition from the "herb form". This diversity has long inspired ecologists in their research, which is why they have been distinguishing more growth forms than just those three I mentioned here. How they arrived at this is quite a long story.

1.1.2 What Do We Know About Evolution?

Spore-bearing plants are currently represented by a reduced range of growth forms. Tree-like forms have survived to the present day only in ferns, in clubmosses, horsetails, and the other spore-bearing plants, we only encounter herbaceous forms. Gymnosperms are all woody plants, and among them, we find only one parasitic non-green plant. The small shrub *Parasitaxus* grows in the dark forests of New Caledonia and uses a fungus to parasitize a member of its own family, a tree named *Falcatifolium taxoides*. Currently, angiosperms are dominant and represent most of the diversity in growth forms, morphologies, and architectures.

Angiosperms were originally small woody plants in the understory of forests composed of gymnosperms (dicotyledonous plants) or they inhabited shallow waters and wetlands (monocotyledonous plants). In evolutionary history, herbs have evolved many times from woody ancestors, as have woody plants from herbaceous ancestors (Klimeš et al. 2022). Although we do not have, and probably will never obtain, definitive proof of why and how this happened, researchers agree that it was the pressure of changing environmental conditions in the Tertiary period, such as the drying and cooling of the climate, that played an important role in the spreading of herbs and establishment of herb dominated ecosystems. Herbs withstand various adverse conditions such as frost, shading, and disturbance better than woody plants. The question is whether the necessary adaptations, which we now know as essential conditions for plant survival in frost (such as narrow vessels, deciduousness, winter dormancy, and frost resistance), developed before the climate cooled and allowed plants to colonize new environments, or whether plants colonized new environments and developed the appropriate adaptations there (Zanne et al. 2014).

While perfect herbs with annual above-ground organs that lack secondary thickening can evolve relatively easily from woody ancestors, from herbaceous ancestors, only relatively unconventional woody plants evolve. The emergence of secondary woodiness from herbaceous forms is associated with special conditions, such as the colonization of isolated islands in the tropical climate zone by herbs (Lens et al. 2013). Some herbs here begin to occupy the vacant niche of woody plants by having their perennial organs grow upwards and form small trunks, instead of growing horizontally on or below the soil surface, or being pulled into the soil by the contractile action of roots (Fig. 1.2).

Also, from monocotyledonous plants, which do not have secondary thickening, woody plants have evolved several times, although their wood is quite atypical (consequently some scholars would not call it wood et all) and only thickens primarily (in palms) or has an unusual method of secondary thickening (*Dracaena*). Nevertheless, they are very successful in the "tree form"; for example, palms in the tropics have the advantage that their crown is made up of leaves, thus not offering much resistance to strong winds, and if their crown is destroyed, it may easily be replaced by the growth of new leaves. Palms, therefore, withstand tropical cyclones (typhoons and hurricanes) better than any other "tree form" species.

1.1.3 Overall Appearance of the Plant

Although the Greek philosopher Theophrastus (Box 2) had already mentioned the division of plants into growth forms in his work, it was Alexander Humboldt (Box 3) who associated ecological significance with them. On his journey through Central and South America, he observed that climatic zones along the altitude gradient are like those along the latitudinal gradient. These analogous climatic zones are then inhabited by plants similar in their overall appearance, physiognomy. The climatological and botanical discoveries he made during his four-year journey were connected to a new scientific field, phytogeography. The slim volume "Ideas for the Physiognomy of Plants" was even one of the first works he published after returning from his travels in 1807. In it, he defined 16 (in the second edition even 19)

Fig. 1.2 Secondary woody plants from the Asteraceae family in alpine communities of tropical climates. (**A**) espeletia (*Espeletia pycnophylla*) around Laguna Voladero (Ecuador, Reserva Ecológica El Ángel); (**B**) coespeletia (*Coespeletia timotensis*) in the area of Pico El Águila (Venezuela, Parque Nacional Sierra de la Culata, both photographs by Petr Sklenář)

categories of the overall appearances of plants. Most categories were taxonomic groups, and aesthetics and inner feeling played a significant role in the "Ideas," as the following quote shows (Humboldt 2011):

> *The verdant carpet which a luxuriant Flora spreads over the surface of the earth is not woven equally in all parts; for while it is most rich and full where, under an ever-cloudless sky, the sun attains its greatest height, it is thin and scanty near the torpid poles, where the quickly-recurring frosts too speedily blight the opening bud or destroy the ripening fruit. Yet everywhere man rejoices in the presence of nourishing plants.*

Humboldt emphasizes that the physiognomy of plants is particularly important for how nature appears in different parts of the world, because plants, due to their large biomass, influence how we perceive nature. He describes individual physiognomic forms of plants (for example: palm form, banana form, cactus form) by mentioning which taxonomic groups they contain and where they occur, thereby giving the described groups ecological significance. For example, about the banana form, he writes the following (Humboldt 2011):

> *In all regions of the earth the palm is found associated with the plantain or banana; (...). This form has a low, succulent, and almost herbaceous stem, the summit of which is crowned with delicately striped, silky, shining leaves of a thin and loose texture. Groves of bananas form the ornament of humid regions; and on their fruit the natives of the torrid zone chiefly depend for subsistence. Like the farinaceous cereals or corn-yielding plants of the north, the banana has accompanied man from the earliest infancy of his civilization. (...) Whilst, however, the cereals, spread by culture over the northern regions, in monotonous and far extending tracts, add but little to the beauty of the landscape; the inhabitant of the tropics, on the other hand, is enabled, by the propagation of the banana, to multiply one of the noblest and most lovely of vegetable productions.*

Humboldt's intuition that the physiognomic forms of plants have some functional relationship to the environment in which they live was inspiring for his successors, who further refined and continued to refine the system of plant growth forms. Some see in growth forms an adaptation of plants to the climate, just like Humboldt, others see an adaptation to soil properties (in this case we shall call those categories life forms, I use this term only for classification by Raunkiær) or simply do not assume an a priori connection with environmental factors and strive only to best describe the diversity of the overall appearances of plants (such categories shall be called growth forms).

1.1.4 Ecological Significance

Eugenius Warming (Box 4), a professor at the University of Copenhagen, one of the founders of ecology and the author of the first textbook, published in 1895 (Warming 1909), considered the significance of the concept of plant growth form for ecology to be comparable to the significance of the concept of species for systematics. For ecology, unlike systematic botany, flowers and fruits are not important according to him, but instead vegetative organs are most responsible for the interaction of the plant with the environment.

Warming developed a hierarchical system of growth forms, which he considered a direct adaptation to the environment. (Table 1.1). His system combines taxonomic group, lifespan, overall appearance, and the habitat where plants grow, which makes the system functional, but the functions of individual forms are not comparable. This problem was solved by Warming's successor in the professorial post at the University of Copenhagen, Christen Raunkiær (Box 5),

Table 1.1 Growth forms according to warming (only main categories)

Hierarchy			Description
I.			Heterotrophic plants (parasites and semi-parasites)
II.			Aquatic plants
III.			Mosses and similar plants (e.g., some small ferns)
IV.			Lichens and similar plants (e.g., some tillandsias)
V.			Liana-like plants
VI.			Other autonomous terrestrial plants
	A		Monocarpic herbs (plants flowering once in their lifetime)
		1	Summer annuals (germinate in spring, then flower and die the following summer)
		2	Winter annuals (germinate in autumn, then flower and die the following summer)
		3	Biennials and perennials (germinate in spring or summer, then flower and die in one of the following years)
	B		Polycarpic perennial plants (live for several years and flower repeatedly)
		1	Perennial herbs (survive unfavorable periods of the year thanks to belowground organs, such as rhizomes)
		2	Rosette plants (live in areas without an unfavorable period of the year and therefore have a constantly present above-ground leaf rosette)
		3	Creeping plants (do not have belowground organs for persistence)
		4	Terrestrial plants with an upright shoot (regardless of where they live, they have constantly present above-ground upright organs, not just a leaf rosette like in category 2. They can be small plants, such as cushion forms, but also medium-sized, like shrubs or stem succulents, or large, like trees

who based his system (of life forms) on one feature, the location of persistent diaspores (buds or seeds) in relation to the soil surface, as a direct adaptation for survival of unfavorable periods of the year, such as drought or frost (Raunkiær 1934) (Fig. 1.3).

Raunkiær's system, unlike many other systems developed over the two centuries since Humboldt's first attempt, is still used in ecology today. The reason is its simplicity. Although only one characteristic (the position of renewal buds/seeds relative to the soil surface) was used in this system, it succeeded in separating classic growth forms (tree, shrub, herb). Thanks to the knowledge of the belowground organs of perennial herbs, Raunkiær was also able to divide herbs according to the depth of renewal buds in the soil or underwater. The entire classification also has a relatively small and therefore memorable number of categories, and all categories refer to one function: survival of unfavorable season. In later years, Raunkiær's system was refined, and additional categories were added to capture the variability of growth forms in non-seasonal climates (compare Fig. 1.3 and Table 1.2).

Raunkiær himself, besides having an excellent knowledge of plants and their belowground organs, was a geographer and statistician, and created the first null model in ecology. In this model, he represented the ideal spectrum of life forms for the world's flora, which he estimated through the analysis of a representative sample

Fig. 1.3 Life forms according to Raunkiær, characterized by the placement of overwintering buds or seeds (**K**) in relation to the soil surface (**D–I**), as well as the water column (**A–C**), and the rooting site (**J**). (**A, B**) hydrophytes; (**C**) helophytes; (**D, E**) geophytes; (**F, G**) hemicryptophytes; (**H**) chamaephytes; (**I**) phanerophytes; (**J**) epiphytes; (**K**) therophytes. Overwintering buds are shown in green, perennial parts of the plant in black, and parts that the plant loses at the end of the season are shown in brown

of species across different biomes. He then related this ideal spectrum to the spectra found in various plant communities. The spectra of life forms for individual biomes are part of biogeography textbooks (Table 1.2). However, today's use of life forms does not utilize this ideal spectrum and is more about a simple comparison of the spectra of different communities. Especially researchers studying vegetation in less explored areas than Europe or North America use life form classification according to Raunkiær as easily ascertainable functional traits of plants (see also further on functional traits).

Although the system of life forms is easy to use, our knowledge of its distribution across the world's flora and vegetation still lags behind taxonomic knowledge. Our awareness is even relatively small regarding the ratio of woody species to herbaceous species in the world's flora, let alone having precise data on the ratio of geophytes to hemicryptophytes. When I ask my students in lectures what they think the ratio between herbs and woody plants in the world's flora is, I receive very different answers. Scientists posed the same question to three groups of respondents. Laypeople estimated an average of 20% woody plants, biology students, as well as professional botanists and ecologists, estimated 30%. In reality, the proportion of woody plants approaches 50% (FitzJohn et al. 2014), and Raunkiær reached the same conclusion a hundred years ago (when we sum up plants in the first five categories for the ideal spectrum in Table 1.2).

Table 1.2 Examples of Raunkiær's life form spectra and the Ideal spectrum taken from Raunkiær's work (1934)

	Number of species	Percentage of species belonging to the life form									
Categories		S	E	MM	M	N	CH	H	G	HH	T
Ideal spectrum	**400**	**1**	**3**	**6**	**17**	**20**	**9**	**27**	**3**	**1**	**13**
Eastern North America											
Baffin Island	199	–	–	–	–	1	30	51	13	3	2
Labrador Coast	246	–	–	2	1	8	17	52	9	5	6
Georgia	717	(0.1)	(0.4)	5	7	11	4	55	4	6	8
Danish West Indies	904	2	1	5	23	30	12	9	3	1	14
Western North America											
St. Lawrence (Alaska)	126	–	–	–	–	–	23	61	11	4	1
Sitka	222	–	–	3	3	5	7	60	10	7	5
Death Valley	294	3	–	–	2	21	7	18	2	5	42
Western Europe											
Franz Josef Land	25	–	–	–	–	–	32	60	9	–	–
Svalbard	110	–	–	–	–	1	22	60	13	2	2
Iceland	329	–	–	–	–	2	13	54	10	10	11
Denmark	1084	–	(0.1)	1	3	3	3	50	11	11	18
Stuttgart	862	–	–	3	3	3	3	54	10	7	17
Others											
Madeira (lowland)	213	–	–	–	1	14	7	24	–	3	51
Egypt (desert)	194	–	–	–	3	9	21	20	4	1	42
Aden	176	1	–	–	7	26	27	19	3	–	17
Seychelles	258	1	3	10	23	24	6	12	3	2	16

S stem succulents, *E* epiphytes, *MM* mega- and mesophanerophytes, *M* microphanerophytes, *N* nanophanerophytes, *Ch* chamaephytes, *H* hemicryptophytes, *G* geophytes, *HH* helophytes and hydrophytes, *T* therophytes

1.1.5 Are We Waiting for the Ideal Classification?

The classifications of growth forms are constantly being developed, and authors usually claim that they are the ones who have come up with the true and best classification. Will we ever arrive at an ideal classification? One problem for an ideal classification is the variability of growth forms, both among species and among individuals of the same species. A single plant species can have different growth forms depending on environmental conditions. Trees at the alpine tree line often form shrubs that are hidden under the snow in winter. In mild winters, nettles overwinter as small green shoots at the soil surface (hemicryptophyte), while in freezing winters, all above-ground parts die, and only belowground rhizomes survive, from which the nettles sprout in spring (geophyte). Some plants have different growth forms depending on their ontogeny. A plant may start its life as a shrub in the forest understory, but later it transforms into a liana, and once it reaches the tree crowns, it may further change into an epiphyte and sever its connection with the soil. Another plant may start its life as an epiphyte, but over time it envelops the trunk of the host tree, suffocates it, and becomes a tree itself.

The problem also lies in the diversity of plant forms on which such a classification can be based. You can use leaves, the woodiness of stems, the placement of regenerative organs, height, or other traits as a basis, and you can also combine them in various ways. We will never have a detailed classification of growth forms for the entire world's flora, but it certainly makes sense to describe growth forms for a particular taxonomic group or in a specific community. In that case, it is possible to develop a classification tailored precisely to the studied system and have categories in the classification that do not occur elsewhere.

A nice example is underground trees in tropical grasslands. The tropical grasslands of Africa and South America are not purely herbaceous communities, but trees and shrubs also occur there, usually adapted to fires. They often have woody tubers (lignotubers) belowground or at the soil surface, from which they can regrow if their above-ground biomass is destroyed by fire. Sometimes these woody plants are not capable of regenerating from belowground organs but are protected from fire by thick bark. Once they successfully reach a greater height, the thick bark protects them from fire, which is fueled by dry grass, and does not reach their crowns. However, there is also a special growth form called the underground tree. It is a tree with a small trunk and branches hidden belowground, from which only the youngest twigs protrude above ground (Fig. 1.4). These trees have evolved from above-ground tree ancestors independently in several families, and if we are studying tropical grasslands, they deserve their special category in the growth form system (Maurin et al. 2014).

Fig. 1.4 Underground tree in tropical grassland in Brazil. Only young twigs are above the soil surface, which burn in a fire, and new twigs grow from the perennial branches stored belowground, where they are protected from fire

1.1.6 Tests of Functioning

Now comes the question of whether we have scientifically tested the function of growth forms correctly and have verified it for use in plant ecology. Of the presented (and unpresented) classifications of growth/life forms, only those based on some testable assumption can be tested (i.e. those that are considered an adaptation to environmental conditions). For example, Raunkiær's classification divides plants according to the position of renewal buds relative to the soil surface and predicts that the position of the buds is an adaptation to adverse seasonal conditions of drought and frost, and that the spectra of growth forms will differ along climatic gradients. The diversity of spectra has been confirmed by repeated observations worldwide. However, it is interesting that there are very few experiments where winter conditions and plant overwintering (or summer conditions and plant "over-summering") with buds placed at various depths or heights have been manipulated.

For experimental verification of the significance of the depth of regenerative organ placement against frost, we must turn to horticultural literature. From experiments examining how best to care for ornamental bulbous plants brought from seasonally dry climatic conditions, we know which of them should be removed from the soil in seasonally cold conditions and let to overwinter in the cellar rather than in the garden, because they would not cope with the frosty winter. For those that can overwinter outside, gardeners must first determine how deep to plant them in the soil. The depth of planting is important not only because bulbs of ornamental plants are better protected from frost when planted deeper, but also because emerging to the soil surface costs them energy (in the form of stored reserve carbohydrates). If we plant them too deep, they are safe over the winter, but in spring they have trouble reaching the soil surface. A shoot growing from a deeply planted bulb may exhaust all its reserves and, moreover, reach the soil surface at a time when other plants have already taken up space. This is, of course, not a problem in a flowerbed where you remove weeds, but in nature, the plant may struggle to accumulate enough reserves to survive the next winter (Lubbe and Henry 2019).

How is a plant able to change (by itself, without the help of gardeners) the position of its overwintering buds? Most seeds germinate on the soil surface, with the stem growing upwards and the root downwards into the soil. Yet all perennial herbs and shrubs eventually have stems (regenerative organs) and thus renewal buds (see also bud bank) belowground. How does this happen? Most often, the root of young plants has a contractile function, begins to shorten, and thus pulls the lower part of the stem into the soil (Pütz and Froebe 1995). However, the plant is not yet victorious; because it grows upwards, the position of the bud bank constantly needs to be adjusted. Perennial herbs that have one root for their entire life may gradually lose the ability to retract into the soil because the root loses its contractile ability, and the renewal buds remain above-ground. This can mean death by freezing for the buds located highest on the perennial belowground parts of the herb, and if the plant does not have other, better-protected buds, it will perish. Perennial herbs that produce new roots from stems throughout their life and shed old roots that no longer have contractile abilities have it easier and can easily maintain the correct belowground position of the buds. Some species can even produce specialized contractile roots differing from other roots, usually species that have tubers or bulbs as overwintering organs (Fig. 1.5).

Another question is: How does a plant know at what depth to store its overwintering organs? We do not know, but it is probably not the cold experienced during winter; it must be sometime during active growth. The plant might use light penetrating the soil surface. For example, plants in an unmown lawn with a thicker layer of litter store their overwintering organs shallower in the soil than plants in a mown meadow where litter is absent. Also, rhizomes of plants growing down or up a slope maintain the same depth below the soil surface. The regulation of organ formation by light is documented by some early experiments with several tuber-forming species (Fig. 1.6), conducted by the Czech botanist and founder of experimental plant morphology, Rudolf Dostál (Box 6), in the first half of the twentieth century. However, we lack experiments comparing multiple species. Dostál's experiments could be an inspiration for ecologists (Dostál 1967). He conducted them to reveal the regulation of organ formation or position by plant hormones. We would need to find out specifically how organs are formed in real ecological conditions and how species differ from each other in this regard.

Further classification of growth forms usually does not assume a direct relationship between form and function, which makes them much harder to test experimentally. Exceptions are experiments concerning individual special categories, which,

Fig. 1.5 Contractile roots in gladiolus (*Gladiolus*), which ensure the placement of renewal buds at the preferred soil depth. The shortening occurs by changing the shape of the cells in the inner layers of the primary cortex of the root and is visible as wrinkles at root surface

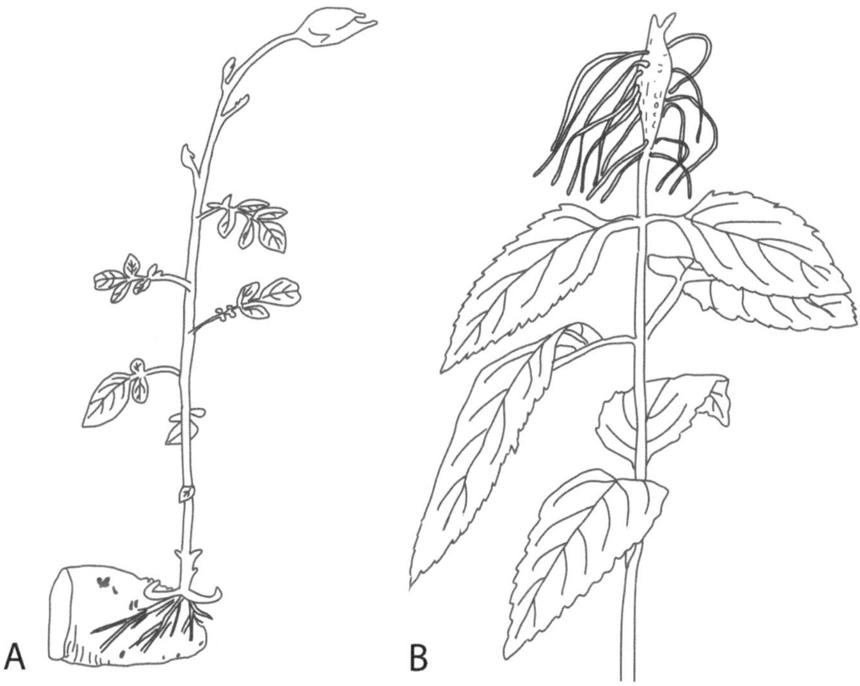

Fig. 1.6 Classic experiments by Rudolf Dostál, in which he manipulated the formation of tubers in potato (*Solanum tuberosum*) (**A**) and *Scrophularia nodosa* (**B**). (**A**) the growth apex of the potato shoot was shaded and formed a tuber; (**B**) the growth apex of the *Scrophularia nodosa* was placed in a moist and dark environment and formed a tuber with adventitious roots. Redrawn according to Dostál (1967)

for example, investigate whether non-green plants really obtain carbon from the host (Julou et al. 2005) or whether lianas save energy when they do not need to invest in the stem (Klimeš et al. 2020). However, these studies concern one specific group extracted from the classification and do not tell us much about how functional the entire system of growth forms is (see lianas and parasites in the classification of growth forms according to Warming in Table 1.1).

References

Dostál R (1967) On integration in plants. Harvard University Press, Cambridge

FitzJohn RG et al (2014) How much of the world is woody? J Ecol 102:1266–1272

Humboldt A (2011) Ideas for a physiognomy of plants. In: Otté E, Bohn H (trans) Views of nature: or contemplations on the sublime phenomena of creation, Cambridge library collection – Earth science. Cambridge University Press, pp 210–352

Julou T, Burghardt B, Gebauer G, Berveiller D, Damesin C, Selosse MA (2005) Mixotrophy in orchids: insights from a comparative study of green individuals and nonphotosynthetic individuals of Cephalanthera damasonium. New Phytol 166:639–653

Klimeš A, Klimešová L, Bartušková A, Klimešová J (2020) Climbing strategy in herbs does not necessarily lead to lower investments into stem biomass. Plant Ecol 221:1159–1166

Klimeš A, Šímová I, Zizka A, Antonelli A, Herben T (2022) The ecological drivers of growth form evolution in flowering plants. J Ecol 110:1525–1536

Lens F, Davin N, Smets E, del Arco M (2013) Insular woodiness on the Canary Islands: a remarkable case of convergent evolution. Int J Plant Sci 174:992–1013

Lubbe FC, Henry HAL (2019) The cost of depth: frost avoidance trade-offs in herbaceous plants. Plant Soil 444:213–224

Maurin O, Davies TJ, Burrows JE, Daru BH, Yessoufou K et al (2014) Savanna fire and the origins of the 'underground forests' of Africa. New Phytol 204:201–214

Pütz N, Froebe HA (1995) A re-evaluation of the mechanism of root contraction in monocotyledons using the example of Arisarum vulgare Targ.-Tozz. (Araceae). Flora 190:285–297

Raunkiær C (1934) The life forms of plants and statistical plant geography. Clarendon Press, Oxford

Warming E (1909) Oecology of plants: an introduction to the study of plant-communities. Clarendon Press, Oxford

Zanne AE, Tank DC, Cornwell WK, Eastman JM, Smith SA et al (2014) Three keys to the radiation of angiosperms into freezing environments. Nature 506:89–92

Comparative Morphology 2

Plant bodies are composed of organs having distinct positions on the plant body, morphological and anatomical characters, and functions. Roots obtain water and nutrients form soil, leaves assimilate carbon using the energy of the sun, and stems connect roots and leaves. Except for those basic organs and their basic functions, plant bodies need to fulfill other functions that are then provided by specialized organs (composed from these basic organs). For example, for resprouting after disturbance, plants employ organs called lignotubers composed from stems and roots deposited belowground that store carbohydrates and the bud bank. In ecology we are using traits of the organs as easy surrogates for complicated functions when comparing strategies across plant species. For example, thinner leaves are more effective in photosynthesis, so by using this measure, we can order plants according to their leaf economic spectrum from acquisitive (thin leaves) to conservative (thick leaves). The function of plant organs approximated by functional traits is tested and used in so called trait-based ecology.

2.1 From Which Parts?

2.1.1 Basic Vegetative Organs

The plant body consists, with few exceptions, of roots, stems, and leaves (I intentionally omit the generative parts of plants here, such as flowers and fruits, as explained in the introduction). From childhood, each of us can distinguish these basic organs (Table 2.1) and we are also aware of their function. Roots anchor the plant in the soil and obtain water and dissolved minerals, leaves assimilate carbon and use it (with the help of light energy) to create organic compounds, which are then transported to the roots via the stem. However, the plant needs much more services from its body parts than it has (basic) organs and can use some of them for other functions. For this new function, the organ usually has a modified shape, or several basic organs participate in fulfilling the function. These shapes on the plant

J. Klimešová, *Plant Bodies*, https://doi.org/10.1007/978-3-031-92192-6_2

Table 2.1 Definition of basic plant organs

Stem	The stem is segmented, has nodes that alternate with internodes, bears leaves at the nodes, and has buds in leaf axils (axillary buds) that can form other leafy stems. The growth of the stem is infinite, or ends with a flower/inflorescence, less often transforms into a thorn, or simply stops growing. Together with the leaves growing from it, a stem forms a shoot. The first shoot of the plant grows from the shoot pole of the embryo (Fig. 2.1), other shoots most often grow from other shoots (axillary shoots), and less frequently from roots (adventitious shoots). Anatomical structure see Fig. 3.7
Root	The root is not segmented, does not bear leaves, usually does not have buds, and its growth is infinite. The first root grows from the root pole of the embryo (Fig. 2.1), other roots grow from roots (lateral roots) or from stems (adventitious roots). Anatomical structure see Fig. 3.7
Leaf	The leaf grows from a node on the stem; its growth is limited, it is usually not segmented, and it does not bear buds. The leaf also does not have an apical meristem, has limited durability, and is mostly flat with branched venation. Together with the stem, it forms a shoot (i.e., a leafy stem).

body, which perform a function other than the basic one mentioned at the beginning, are called derived organs or metamorphoses. For example, a potato serves the plant for vegetative reproduction and overwintering; it is essentially a stem specialized for a new function; this special derived organ or metamorphosis is then called a stem tuber.

The establishment of the definition of basic organs has changed throughout history, both according to the observational abilities of the researcher, and according to technological progress. Anatomical features, of course, entered the definitions only after the invention of the microscope. We can illustrate the development in understanding basic organs by what the Greek philosopher Theophrastus writes about roots. He begins by stating that the plant has roots below ground. The belowground position of roots was thus part of the definition of this organ at that time. However, he points out that in some plants, belowground organs have the properties of stem, and the plant is capable of growing and reproducing from them. He concludes from this that not all organs that are below ground should be called roots. However, if we insist on using the term roots, he points out, then we must distinguish between two types of roots: those that serve for nutrition and those that serve another purpose. He simultaneously mentions the aerial roots of the fig and ivy. In the fig, roots grow from the branches to the ground, forming additional support for its large crown, and eventually creating a small grove; in ivy, the roots assist in climbing other plants by growing into surface irregularities, mainly on the bark of trees, which they climb.

Basic organs are distinguished by their position relative to other organs, development in ontogeny, and morphological and anatomical structure. In the picture (Fig. 2.1), we see a young dicotyledonous plant consisting of a shoot growing upwards and a root growing down into the soil. These two growth directions are already established in the plant embryo, which has a shoot and root pole. The shoot further consists of a stem, which is segmented (modular), and leaves growing from nodes. On this plant, we can further distinguish a module at the base of the shoot, which has a transitional anatomical structure between the stem and the root, called

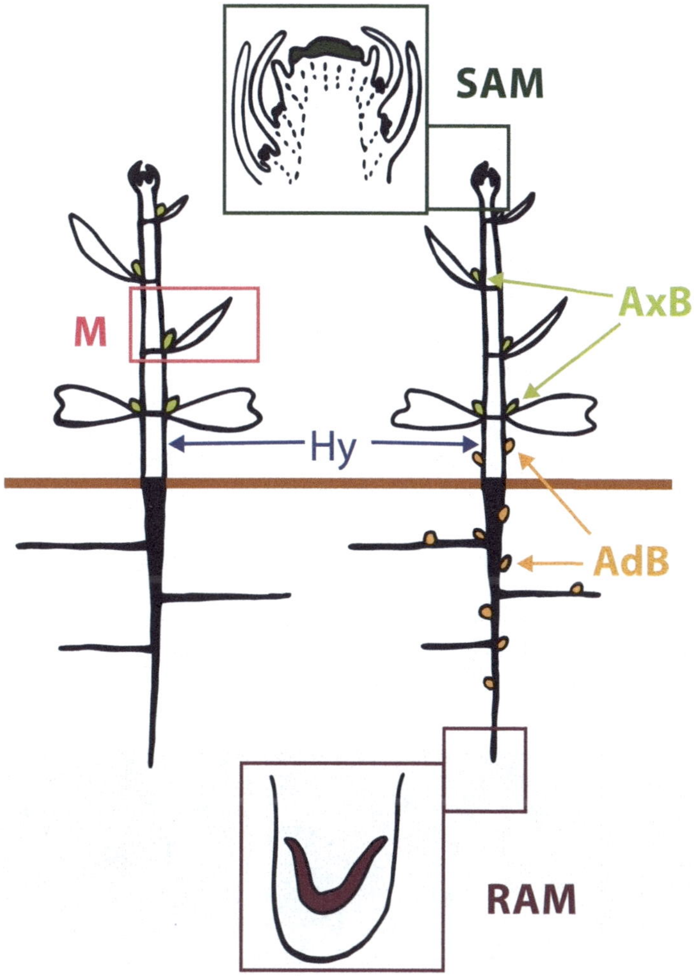

Fig. 2.1 Diagram of a dicotyledonous seedling. M the smallest repeating unit of the shoot, a module, consisting of an internode, node with leaf, and axillary bud, SAM stem apical meristem, RAM root apical meristem, AxB axillary buds, AdB adventitious buds, Hy hypocotyl. On the left is a plant that is unable to produce adventitious buds, and on the right is a plant that has this ability

the hypocotyl. We also see growth tips, meristems, at the top of the shoot and root (apical meristems), and others in the axils of leaves (axillary meristems). This is the structure of a young angiosperm plant, its shoot is called the primary shoot, and its root the primary root. Most plants continue to grow by branching and producing additional shoots and roots (see the chapter on architecture). The morphological diversity observed in nature is partly due to different shapes (thickening, shortening, and elongation), reduction, multiplication, or fusion of basic organs, and partly due to different arrangements and frequencies of branching. The mutual position of organs relative to each other and their anatomical structure is usually preserved, allowing us to use them to identify derived organs.

You may not find all parts of a plant at any time of the year (for example, leaves are missing on deciduous trees in winter). More importantly, however, there are plants that have lost one of the basic organs during evolution, or have the organ, but it serves a completely different function than the basic and typical one. This mostly concerns leaves, which in cacti, for example, are transformed into spines, or roots, which are completely absent from aquatic carnivorous plants of the genus Utricularia; losing a stem is rarer (see architectural peculiarities). The role of such evolutionarily lost basic organs is taken over by another organ, otherwise, the plant would not survive. In cacti, the stem often assimilates carbon, and in aquatic carnivorous plants, leaves take up nutrients instead of roots. For example, the carnivorous plant *Genlisea* has two types of leaves, some are large, green, and assimilate carbon, others are white, located belowground, and fused into a tube that serves as a trap for small creatures wandering in the moss (Fig. 2.2). The belowground leaves, called rhizophylls, anchor the plant in the substrate and obtain nutrients, thus fulfilling the role of roots.

Organs can also be lost due to disturbance. For example, plowing cuts weed plants into fragments that survive because they can create missing organs. Stems are the best at surviving fragmentation; if the fragment contains at least one bud, the other basic organs can easily be restored. As we know from the definition, if only a leaf or just a root, which usually do not bear any buds, remains from the whole

Fig. 2.2 Rootless carnivorous plant *Genlisea hispidula* (Lentibulariaceae). (**A**) overall view of the belowground part of the plant growing in peat; (**B**) two types of leaves: green ones assimilate above the soil surface and white ones catch prey below the soil surface; (**C**) detail of the belowground leaf; (**D**) detail of the flower; Tr trap (rhizophyll), TA trap arms, Ve vesicle for digesting prey. (Photo Lubomír Adamec)

plant, then regeneration is not possible. However, nature can help itself because roots, and less often leaves of some species, can also create buds under certain conditions and then regenerate. In our flora, about 10% of plants can make buds on roots (Bartušková et al. 2017); only a few species can make buds on leaves (Fig. 2.3). In the floras of other regions, however, we can expect different ratios, for example, the leaves of trees of tropical rainforests commonly regenerate into small plants.

Besides the basic functions (leaf = assimilation, root = absorption, stem = connection), the basic organs also ensure other functions. We mentioned, for example, the regenerative function of the stem, which makes the stem considered to be a vital organ. Other examples can be defensive functions (various thorns), clonal growth and overwintering (rhizomes and bulbs), and so on. The contribution of the root, stem, and leaf to these functions can vary among different species, and moreover, the root, stem, and leaf can combine to form a single derived organ. This morphologically complex situation is a paradise for morphologists, but is it also interesting for ecologists? Does it matter to an ecologist whether a thorn is of stem, root, or leaf origin? It does matter because they are interested in what evolutionary constraints accompany the emergence of a particular organ and thus its function. Moreover, what if a leaf thorn has different properties than a stem thorn? For example, it has been found in African savannas that thorniness as an adaptation to reduce grazing has evolved at least 50 times independently in different plant groups. Although thorns can develop from leaves, twigs, or surface tissues of the stem, no one paid

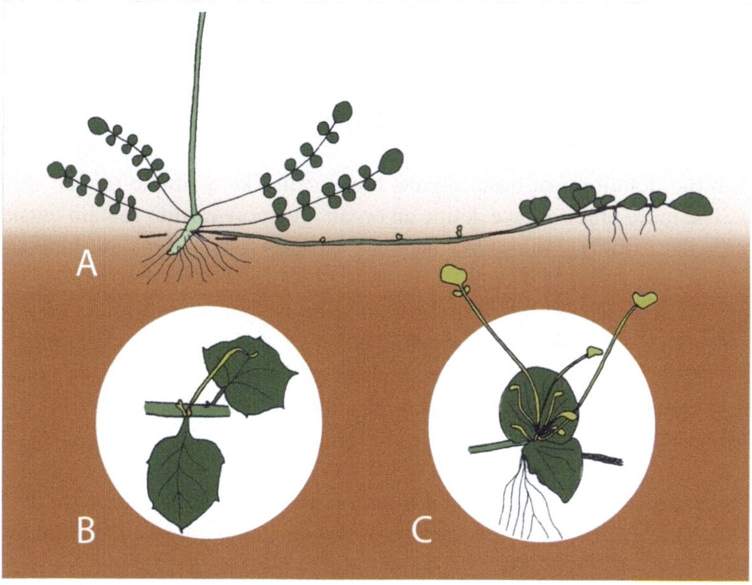

Fig. 2.3 *Cardamine pratensis* is the only plant in the flora of Central Europe capable of spontaneously producing adventitious buds on leaves without disturbance. (**A**) lower leaves of the basal rosette elongate after the plant flowers and begin to root; (**B**) clusters of buds appear between the leaflets of the compound leaf; (**C**) some of the buds grow into small plants with their own roots

attention to this until recently. That was a mistake, because different types of thorns develop at different stages of plant ontogeny depending on their morphological origin: young plants with thorns formed by twigs can only defend themselves much later than those that form thorns on leaves. This is because leaves are formed by seedlings much earlier than branches. This difference can lead to varying success of differently thorned plants in establishing from seeds in heavily grazed communities (Armani et al. 2019).

2.1.2 What Do We Know About Evolution?

The differentiation of the plant body into roots, stems, and leaves had already occurred in spore-bearing plants, so all seed plants have these organs differentiated. Non-vascular spore-bearing plants (mosses) have stem-like, root-like, and leaf-like structures differentiated, but they are not true roots and leaves. These developed repeatedly in parallel with the emergence of the vascular system of vascular spore-bearing plants through the specialization and modification of stems. The first roots were branches of belowground stems that specialized in geotropic growth (Kenrick and Strullu-Derrien 2014). Leaves, on the other hand, are derived from specialized branches of the above-ground parts of stems, which soon ceased their longitudinal growth but grew in width, or several of them fused to form a flat leaf organ (Tomescu 2009). This happened in the Devonian (416 to 360 million years ago), and tree-like forms of spore-bearing plants already had an extensive and diversified root system and a great variety of leaf shapes.

2.1.3 Classical and Continuous Model

Although the definition of basic organs may seem like a simple matter, it has its pitfalls. The more precisely we define an organ, the more rarities remain outside the categories because the boundaries between basic organs are not as sharp as they might seem at first glance. For this reason, the history of plant morphology is lined with various approaches and classifications. For our purposes, I select two: (i) the classical model, which has clear boundaries between categories and classifies transitional types as oddities, and (ii) the continuous model, which allows for diffuse boundaries between categories and classifies transitional types into a continuum of forms (Fig. 2.4).

(i) A representative of the classical model is the German morphologist Wilhelm Troll (Box 6). In his multi-volume work on comparative plant morphology (Troll 1941–42), he dedicates one volume to each organ. He defines the given organ and provides a list of examples of its morphological diversity, including modifications (metamorphoses, see further) for other functions. In this system, transitional forms have a place among curiosities. For example, in lists of teratological phenomena (atavisms), there is list of species in which developmental

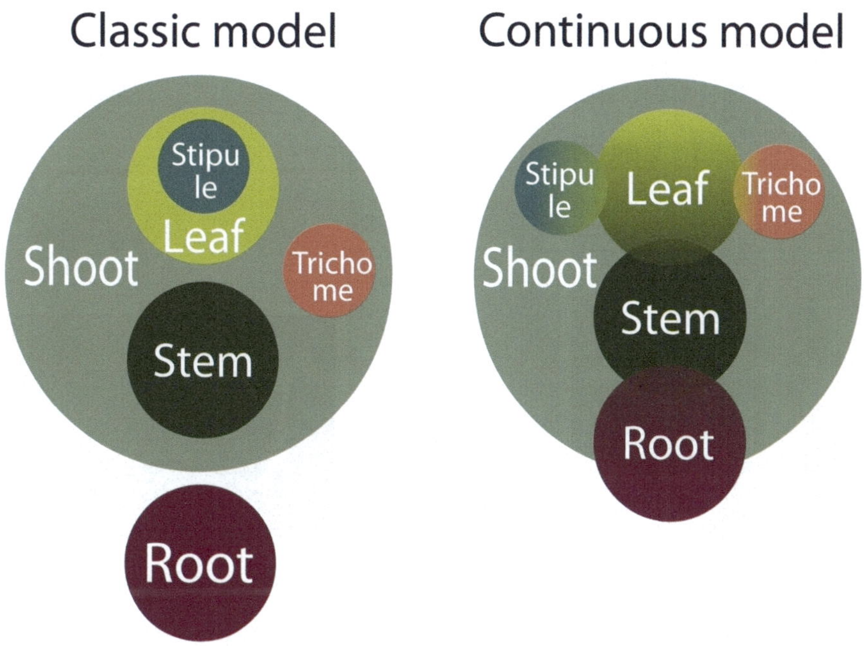

Fig. 2.4 Diagram of the classic morphological model with clear boundaries between individual categories (left) and the continuous model, which allows for transitional forms (right)

anomalies were observed, such as the multiplication of floral parts, the production of buds on roots, etc. (Penzig 1921–2).

(ii) We consider the English botanist Agnes Arber (Box 7) as a representative of the continuous model. She also wrote several morphological books, but her approach differed. She always worked on some ecological (aquatic plants (Arber 1920)) or taxonomic group (grasses (Arber 1934)) and recorded the diversity of morphological forms within these groups. She did not focus so much on the questions of precise and all-encompassing classification; instead, she paid attention to transitional forms. Transitional forms did not represent an oddity for her but a potential that an organ has in its development but is not always expressed. One of her conclusions, for example, is that a leaf tends to take on the character of a stem, i.e., in some plants, a leaf can have infinite growth or can bear buds.

Although Wilhelm Troll was valued in his time for his precise systematic approach, nowadays, when plant morphology deals with issues other than precise classification (such as hormonal control of organ development), Agnes Arber's approach is being rediscovered (Kirchoff 2001). It is beneficial not only for developmental biologists but also for ecologists. After all, the fact that many plants are capable of creating buds on otherwise budless hypocotyls or roots after disturbance (Fig. 2.5), regenerating from these buds, and completing their life cycle, is not insignificant for

Fig. 2.5 Adventitious buds and shoots on the hypocotyl (**A, C**) and roots (**D, E**). (**A, B, C**) *Euphorbia lathyris*, (**D**) *Centaurea scabiosa*, (**E**) *Trifolium rubens*, Hy hypocotyl, AxB axillary buds, AdB adventitious buds. Note the difference between axillary buds (**B**) growing from the node area on the stem in the leaf axils and adventitious buds (**C**) scattered irregularly on the hypocotyl

the population dynamics of the plant. It also seems that it is not entirely by chance that plants capable of sprouting from roots most vigorously are found especially in heavily disturbed habitats (Bartušková et al. 2021). Why then consider the formation of buds on roots as some insignificant anomaly? Rather, we could consider the ability to form buds after disturbance as another possible function of roots, a regenerative function.

2.1.4 Archetypal Plant

Interestingly, both researchers, Agnes Arber and Wilhelm Troll, were inspired by the same source, admiring the founder of the field, Johann Wolfgang Goethe (Box 8), and his idealistic search for the archetypal form of the plant (Urpflanze = archetypal plant). Goethe came up with the idea that the diversity in plant forms we observe is caused by the metamorphoses of an ideal archetypal form of a plant (see also the chapter on Architecture).

How did the world-renowned poet Goethe come to the issue of metamorphosis in plants? It was the second half of the eighteenth century, and the natural sciences were experiencing unprecedented development, for example, thanks to the systematic work of Linnaeus (Box 9) and exploratory journeys that filled emerging botanical gardens and herbaria. Goethe was interested in progress in the natural sciences and admired the new Linnaean system of plant classification, but at the same time, he was dissatisfied with its complexity. In 1790, he wrote a slim work

"An Attempt to Explain the Metamorphosis of Plants" (Goethe 1790), which laid the foundation for the field of comparative plant morphology. Strictly speaking, it is not a classic scientific communication but rather an essay. Although he drew from and built upon previous researchers, there is not a single citation in this essay. Nevertheless, it was a very influential work that some condemned and others loved.

Goethe's idea was simple and could be summarized in one sentence: everything is a leaf (including all floral organs). The perianth, corolla, calyx, stamens, and pistils are all organs of leaf origin (Fig. 2.6). The evidence for this claim was various plant atavisms, i.e., forms where leaf parts had a transitional appearance between a green assimilating leaf and a floral part. Goethe's followers then developed this textual idea and supported it with further evidence. They expanded the idea to all parts of the plant body and searched for homologies (organs of the same origin but different functions) and analogies (organs of the same function but different morphological origins (Owen 1848)). It is interesting that today, floral parts are no longer considered transformations of leaves; according to the latest research, the entire flower seems to be more homologous with the sporangium of spore-bearing plants (Classen-Bockhoff 2016).

2.1.5 Metamorphoses

Transformations (metamorphoses) of basic organs to perform functions other than their main ones (photosynthetic for leaves, transport for stems, and absorptive for roots) are taught to children in elementary school, yet we will briefly review them here.

A leaf can transform into a tendril, storage or protective scale, spine, or diaspore. Leaf tendrils are found in leguminous plants, where only the upper leaflet of a compound leaf, several leaflets, or all can be transformed into a tendril, and photosynthesis is then ensured by an enlarged stipule. This transformation ensures the plant's attachment to a support, and its bearers parasitize other plants by utilizing their structures, hence they are called structural parasites. Storage leaves are the most commonly found in plants with bulbs—a bulb is composed either of storage scales, as in garlic, or storage bases of leaves, as in the kitchen onion. In various species of garlic, lilies, snowflakes, snowdrops, and others, we find various combinations of storage scales, storage bases of leaves, and quite ordinary leaves. Storage leaves are also the main mass of overwintering bodies (serving also for clonal multiplication) of aquatic plants, which we call turions. A protective scale is on every dormant bud; on one that has never produced green leaves, but also on one that formed normal green leaves and switched in winter to create several small and hard scales protecting the growth apex from frost. Leaves of some species are not green but colored and, together with flowers, form what is called a biological flower, something we perceive as a flower but is actually an inflorescence, in which leaves also participate. Such inflorescences are characterized by their great durability and longevity, so they are often used as decorations, for example, the inflorescence of hydrangea. A leaf

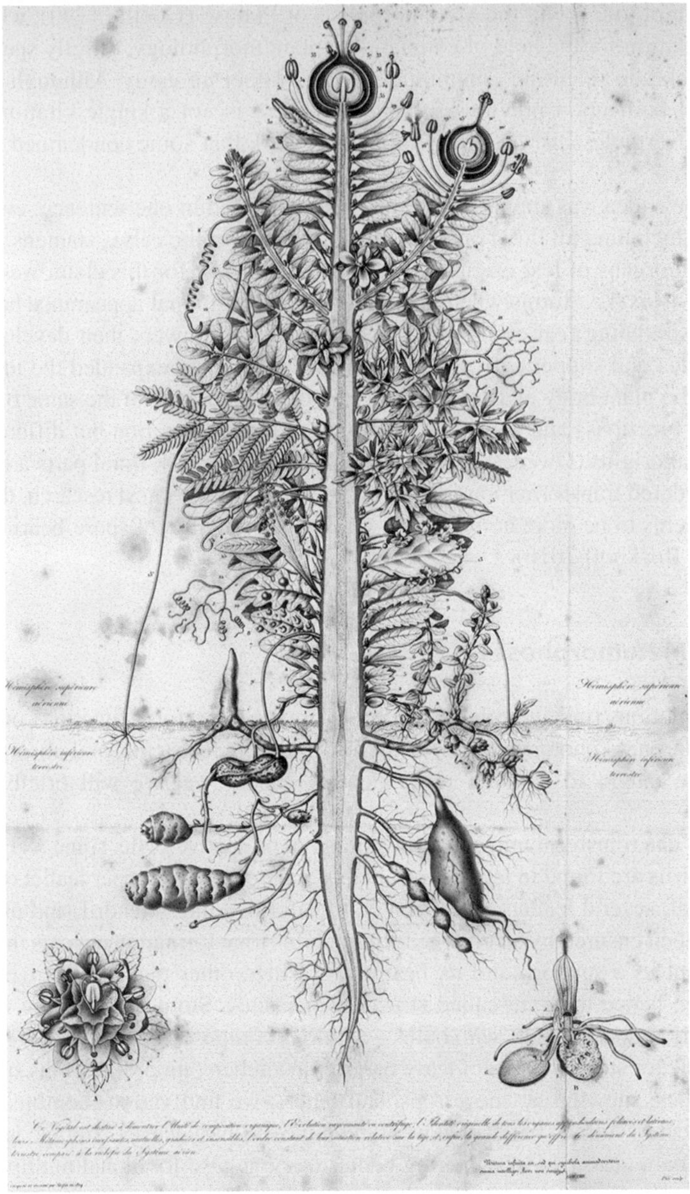

Fig. 2.6 The archetypal plant as depicted by the French botanist Pierre Jean Turpin at Goethe's request. (Wikimedia, Public Domain)

can also be a thorn, as in cacti, or, usually after breaking off, can serve for clonal multiplication in some leaf succulents. For clonal multiplication, the leaf needs not only reserves but also a bud—this is either an axillary bud of the stem from which the leaf fell, embedded in the leaf tissue, or it arises adventitiously directly on the leaf (Fig. 2.7).

Fig. 2.7 Leaf metamorphoses. (**A**) tendril in *Lathyrus*; (**B**) storage scale in garlic (*Allium*); (**C**) turion in *Utricularia macrorhiza* (photo by Lubomír Adamec); (**D**) protective scales of overwintering buds in *Aesculus hippocastanum*, (**E**) colorful leaves in *Hydrangea* inflorescence; (**F**) spines in *Gymnocalycium*, (**G**) funnel-shaped leaf trapping prey in carnivorous pitcher plant (*Sarracenia leucophylla*), (**H**) leaf as a reproductive organ in succulent plant *Kalanchoe bryophyllum*

Stem transformations are no less rich. A stem can be storage, as in bottle trees and cacti, or twining as in vines. The stem forms a spine (thorn) in blackthorn (*Prunus spinosa*) or takes on the appearance and function of a leaf, as in butcher's broom (*Ruscus*). Stems also serve for vegetative reproduction and transform into runners and rhizomes (see below). When they combine their storage function and belowground placement, we speak of tuberous rhizomes or stem tubers (Fig. 2.8).

Root transformations are also remarkable. They include transformation into an attachment in some vines, such as *Hedera helix* or *Parthenocissus quinquefolia*, transformation into thorns in tropical trees, transformation into storage, supportive, respiratory, and reproductive roots. We know storage roots from the kitchen from carrots, parsley, or beets, and root tubers are found in sweet potatoes (*Ipomoea batatas*) or cassava (*Manihot esculenta*), but they also occur in wild plants, such as lesser celandine (*Ranunculus ficaria*) or orchids (e.g., *Dactylorhiza majalis*). Breathing roots can be found in swamp trees of the tropics and subtropics, as well as the various supportive roots that provide support for giant trees in the soft soil of the tropical rainforest. Many species have roots capable of forming buds (see also chapter 2.1.1), and such species are very successful weeds of arable land (*Rumex acetosella*; Fig. 2.9).

Surely, I have omitted some of your favorite metamorphoses in previous lists, but you must realize that Wilhelm Troll wrote an entire book about each basic organ! I'll add at least a small bonus: if you like radishes, then you surely know that it is a hypocotyl tuber. However, hypocotyl can also have regenerative abilities and form buds, as is the case with flax (*Linum usitatissimum*). Inflorescences

Fig. 2.8 Stem metamorphoses. (**A**) Storage and photosynthesizing stem in the saguaro cactus in the Arizona desert (*Carnegiea gigantea*, photo by Jiří Doležal); (**B**) twining stem of a herbaceous vine *Ipomoea purpurea*, (**C**) stem thorns called spines in *Prunus spinosa*; (**D**) photosynthesizing branches shaped like leaves in *Ruscus aesculentus* (photo by Jana Navrátilová); (**E**) hypogeogenous rhizome in *Elytrigia repens*; (**F**) epigeogenous rhizome in *Filipendula ulmaria*; (**G**) runner in strawberry (*Fragaria vesca*); (**H**) stem tuber in potato (*Solanum tuberosum*)

Fig. 2.9 Root metamorphoses. (**A**) adventitious roots in *Hedera helix*, (**B**) supportive roots in *Ficus bengalensis*, (**C**) breathing roots in mangrove tree (*Avicennia*) from southern Taiwan, (**D**) storage taproot in carrot, (**E**) storage adventitious root in *Ranunculus ficaria*, (**F**) roots serving as clonal growth organ in *Geranium sanguineum*, (**G**) root serving as clonal growth organ in *Rumex acetosella*

(which we do not mention much here) can serve not only for seed reproduction but can transform into an organ of clonal growth and produce offspring asexually. It usually appears that instead of some flowers, there are small plants, tubercules, or bulbils that did not arise from seed germination, as it might seem at first glance, but grew from axillary buds. This phenomenon is called pseudovivipary

Fig. 2.10 Thorns of different morphological origins. (**A**) basic shoot organs: Ste. stem, Le leaf, Sti stipule, AxB axillary bud; (**B**) thorns originating from the surface tissues of the stem in *Rosa*; (**C**) thorns originating from stipules in *Robinia pseudoaccacia*; (**D**) thorns originating from leaves in *Ferocactus coloratus*; (**E**) thorns originating from axillary branches in *Prunus*

and is often found in alpine and arctic species (for example, in alpine bluegrass, *Poa alpina*).

The previous examples were homologies, organs of the same origin serving different functions. If we wanted to discuss analogies (different organs adapted to serve the same function), it would suffice to arrange the examples differently. For instance, the organs lianas are using to ensure attachment to their support host may be formed by tendrils from different leaf parts (*Vicia*), hooked hairs (*Galium aparine*), twining stems (*Calystegia sepium*), or aerial rootlets with matted pads (*Hedera helix*). These organs differ so much that they determine which support their owners can climb. While plants with rootlets may climb old tree trunks, tendrils allow plants to support themselves using neighbouring herbs, and twining stems specialize on thin woody trunks. Another example are thorns that may originate from entire leaves (*Ferocactus coloratus*), parts of leaves (*Robinia acacia*), stems (*Prunus spinosa*), and from the surface tissues of stems (*Rosa*) (Fig. 2.10).

2.2 What Is the Function of Basic Organs?

2.2.1 Functions and Strategies

When we talk about the function of an organ, it is not because the organ has it written in a user manual. In reality, understanding the functions of basic plant organs is still ongoing. Although Theophrastus knew that leaves are exposed to the sun, he had no idea what they do with that sun. Today, we have very detailed information for some functions of plant organs, such as photosynthesis, while for other functions, we are still groping. What interests ecologists, however, are not just physiological processes, but whether and how the performance of a relatively complex function is related to some easily measurable properties of the organ that would reflect how the performance of the given function differs among species. With this information, ecologists could then, for example, test how the given function affects the distribution of species in the landscape.

A citation about the ecological significance of leaf size and persistence from Humboldt's work on the physiognomy of plants can be used as evidence that botanists have long noticed the relationship between simple properties of a plant organ and its function (Humboldt 2011): "*While the Banana form presents us with the greatest degree of expansion, the Casuarinas and the acicular-leaved trees exhibit the greatest contraction of the leaf-vessels. Pines, Thujas, and Cypresses constitute a northern form but rarely met with in the tropics (…). Their evergreen foliage enlivens the gloom of the dreary winter landscape, while it proclaims to the natives of the polar regions that, although snow and ice cover the surface, the inner life of plants, like the Promethean fire, is never wholly extinct on our planet.*"

As the number of botanists increased through history, so did the information about plants they accumulated, and gradually efforts began to appear to organize and synthesize this information. From the end of the nineteenth century, so-called ecological (or biological) floras were published. These are compendiums where information is gathered for individual plant species about their environmental requirements (soil, climate), occurrence with other plants, geographical distribution, life cycle, seed production, biomass, germination, herbivores, diseases, and pests. From the beginning of the century until the Second World War, for example, a series of books titled "Life History of Plants of Central Europe" was published, in which several hundred species were described (Kirchner, O., Loew, E. Schröter, C. [eds.] Lebensgeschichte der Blütenpflanzen Mitteleuropas). Additional species were and still are described in series of articles dedicated to certain geographical areas, for example, the biological flora of the British Isles (published in *Journal of Ecology*), the ecological flora of the Moscow region (a collection of proceedings published by Moscow University), the biological flora of central Europe (published in the journals *Flora* and *Perspectives in Plant Ecology, Evolution and Systematics*), or to a certain type of vegetation (for example, the Biological Flora of Canadian Weeds, Published in the *Canadian Journal of Plant Science*). In the 1970s and 1980s, supplementing information on the ecology of individual plant species became very popular. The most active and influential at that time was the group of comparative ecology led by Philip Grime (Box 10). This group opposed the random collection of all possible data about plant life in biological floras with systematic measurement of sets of traits under standard experimental conditions (Grime et al. 2007).

Philip Grime, based on his research, defined three extreme plant strategies (Grime 1977), that is, groups where plants exhibit similar properties and similar behavior in different ecological situations (Fig. 2.11): (i) Competitive strategy (C strategists)—in conditions without disturbance and stress, competitively capable tall plants are most successful; (ii) Stress-tolerant strategy (S strategists)—in

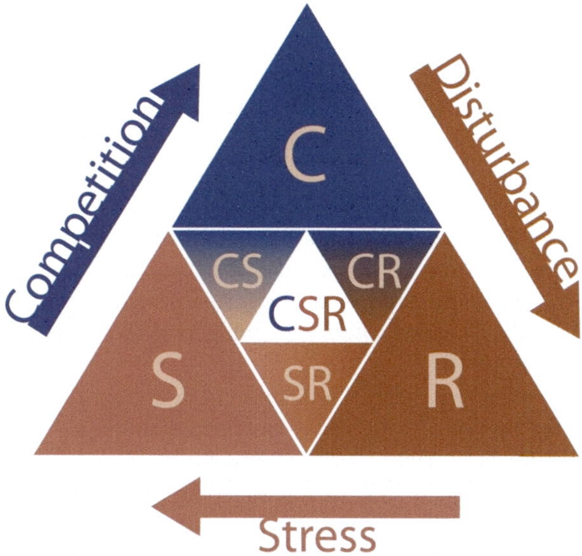

Fig. 2.11 Strategies according to Grime. C—representatives of the competitive strategy dominate in conditions where there are favorable living conditions and no disturbance; R—plants with a ruderal strategy dominate in habitats with favorable growth conditions but frequent strong disturbance; S—stress-tolerant species dominate vegetation where there are few resources, but also no disturbance. The combination of letters indicates transitional strategies along gradients of nutrients and disturbance. Grime explains why there are three strategies and not four, as would correspond to the combination of conditions (abundance and scarcity of nutrients versus disturbance and non-disturbance), due to the impossibility for plants adapt to both disturbance and nutrient deficiency

conditions where growth is limited by some factor such as nutrient deficiency, cold, or too much water, slow-growing species investing in special adaptations that help them survive stress are successful; and (iii) Ruderal strategy (R strategists)—where some disturbance repeatedly removes a large amount of produced biomass, but there are enough resources for growth, plants characterized by rapid growth and high production of small and easily dispersible seeds dominate. R strategists manage to complete their life cycle between two disturbances, and easily dispersible seeds have a chance to find a new freshly disturbed place without competition from other plants.

Grime's classification of strategies (Grime 1977) was and still is very influential and inspiring. However, it can be frustrating for researchers from other parts of the world than the global north: that they do not have such detailed information for their floras that would allow them to use the same system of strategies; that the position of plants in the triangle is relative and should change with the addition of each new species; that it only considers the strategies of adult plants, even though the strategies of juvenile plants may differ; that the system is categorical and difficult to calculate with; that it imperfectly describes adaptations to disturbance because it does not consider disturbance tolerance and plant regeneration from belowground organs; that determining a plant's strategy requires rich field experience, and thus it is often a very subjective opinion as to how classify a particular plant.

At the end of the 1990s, ecologists formulated, in two independently published articles, the idea of using easily measurable plant traits to estimate the vital functions of the plant. The innovation of this approach lies in the fact that it was meant to replace complex and often subjective categorization with simple repeatable measurements. The idea of how to formalize the mutual comparison of ecological strategies of more plant species, was a significant advancement from subjectivity to objectivity. At the same time, it was possible to expand the geographical scope of comparative ecology beyond Europe and North America, into less explored types of vegetation, and involve students who were beginners in the field and lacked extensive field experience.

The two independently published articles came from two research groups. The group of Northern European and North American authors, led by the American Evan Weiher (Weiher et al. 1999) (let's call this the northern approach), identified the basic functions of a plant such as reproduction, persistence, and competitive ability (Table 2.2) and found traits responsible for these functions. They proposed that measuring simple morphological traits could be used to estimate these functions.

Table 2.2 Challenges plants face and the traits they use to address them

Challenge		Complex trait	Simple trait (examples)
Dispersal	In space	Dispersal distance	Seed mass, seed dispersal method
	In time	Seed persistence in the seed bank	Seed mass & shape
Establishment	Seedling growth	Seed mass, relative growth rate	Seed mass, SLA, LWC
Persistence	Seed production	Fitness	Seed mass, aboveground biomass
	Competitive ability	Response to competition	Plant height, aboveground biomass
	Plasticity	Growth response to stress (and its dependence on ontogeny)	SLA, LWC
	Space occupation/plant persistence	Lifespan	Growth form, stem density
	Space occupation	Vegetative spread	Clonality
	Response to disturbance	Tillering, palatability	Ability to tiller after disturbance
	Response to stress	Phenology	Start of flowering, SLA, LWC

According to Weiher et al. (1999). *SLA* specific leaf area (area of one gram of leaf); LWC *leaf water content*

Table 2.3 What simple traits in plants are recommended to study by the northern (all) and southern working groups (only Seed weight, SLA, and Plant height)

Trait	Function
Seed weight	Dispersal distance, persistence in seed bank, success in establishment, fertility
Seed shape	Persistence in seed bank
Seed dispersal method	Dispersal distance, persistence in seed bank
Clonality	Space occupation
LWC, SLA	RGR, plasticity, stress tolerance ability, evergreeness, leaf persistence
Plant height	Competitive ability
Aboveground biomass	Competitive ability, fitness
Life history	Plant persistence, ability to maintain space, disturbance tolerance
Flowering onset	Stress avoidance, disturbance avoidance
Resprouting after disturbance	Disturbance tolerance
Stem density	Plant persistence, carbon storage

According to Weiher et al. (1999)
LWC leaf water content, *SLA* specific leaf area, *RGR* relative growth rate

The southern approach, represented by the Australian researcher Mark Westoby (Westoby 1998), goes even further and states that most plant functions are correlated, so we can, in line with Grime, distinguish several plant strategies. To characterize strategies, we do not need to measure many traits, but only the essential ones. As representative for understanding plant strategy, it is sufficient to measure plant height, which tells us how competitive the plant is (C strategists from Grime's triangle excel in this), specific leaf area, which tells us how efficient photosynthesis is (low due to nutrient deficiency for S strategists), and seed size, which reflects the maternal plant's investment in the offspring, its dispersal possibilities, and its ability to establish in a competitive environment (low for R strategists). Both approaches, northern and southern, overlap precisely in these three recommended traits, and functional ecologists have focused, for another 20 years since the publication of the cited articles, on measuring (mainly) those traits for as many species as possible from around the world (Table 2.3).

2.2.2 Easily Measurable Traits

The trio of Westoby's plant traits, known as LHS traits (from "Leaf, Height, Seed"), allow for precise classification of plants into the plant strategy system, i.e. into the LHS scheme. The LHS scheme is intended to: describe meaningful variability in the ecological traits of plants, classify any plants growing anywhere in the world, and ensure easy measurability. Furthermore, these key traits should also exhibit less intraspecific than interspecific variability and should be independent of each other.

SLA, specific leaf area, tells us what area a unit biomass of the plant's photosynthetic apparatus has. A large specific leaf area means greater photosynthetic efficiency of the leaf, greater demand for nitrogen and water supply, faster growth and shorter lifespan, greater palatability for herbivores, and easier decomposition of dead biomass (for example, nettle). Plants with smaller SLA exhibit opposite traits (for example, rhododendron). When we rank plants according to SLA, we obtain an economic spectrum of the leaf from conservative, with slow growth in poor environmental conditions, to acquisitive (focused on acquiring new resources and not on their recycling), with rapid growth in resource-rich environments (Fig. 2.12).

The potential height of a plant's foliage in adulthood indicates the plant's competitive abilities. Competition for light among plants is asymmetrical because a larger plant not only gains more light than a smaller plant but also shades the smaller plant with its foliage, thereby gaining a disproportionate advantage. However, it is not just the position of the plant's leaves in the dense vegetation that determines how much light, heat, and moisture the plant receives, but also how much it needs to invest in supportive structures to maintain its height, and in seeds so that young plants can succeed in the race for light. The offspring of tall plants must either be able to establish themselves in the shade of adult plants and wait for a natural disaster that destroys the vegetation or find a place where the vegetation is destroyed. In such places, succession (the development of plant cover) occurs, where the race for light starts anew: initially, even small plants can thrive, but gradually larger and more persistent plants grow or appear, overshadowing the small plants. Thus, the height of a plant's foliage also indicates the plant's position in the succession of vegetation after its large-scale destruction (Fig. 2.13).

What is the relationship between plant height and plant organs? Small plants, herbs, can reach their maximum height with large leaves, leaves on long petioles, or tall stems bearing leaves. In seasonal climates, they start their race for light anew each growing season, and their height is therefore limited. In the tropics, the size of herbs is instead limited by the mechanical properties of the stem, which cannot compete in height with woody forms. Woody plants, shrubs, and trees, achieve their height through the growth and persistence of sturdy stems (Fig. 2.14).

Fig. 2.12 The leaf economic spectrum, which classifies plants based on leaf traits along a resource availability gradient. Plants with leaves that manage resources cautiously (conservative) are typical for resource-poor sites, while plants with leaves that do not conserve resources (acquisitive) are typical for resource-rich sites. SLA specific leaf area, N nitrogen, P phosphorus

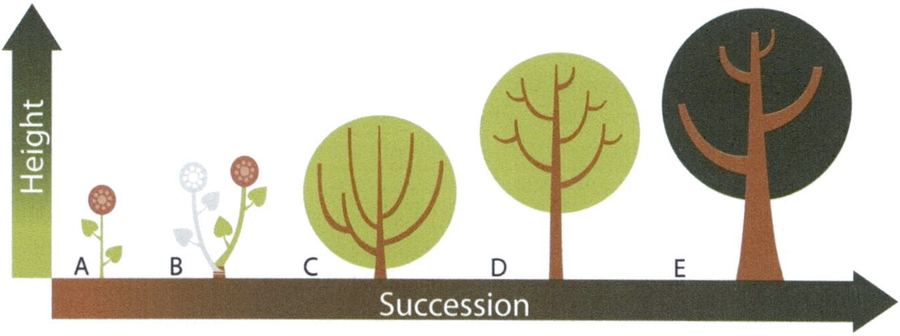

Fig. 2.13 During succession, small plants are gradually replaced by larger plants, the faster-growing ones are replaced by those which grow more slowly but reach a larger size. (**A**) annual, (**B**) perennial, (**C**) shrub, (**D**) fast-growing tree, (**E**) slow-growing tree

Fig. 2.14 How do plants achieve different heights? (**A**) The Arctic tundra in Svalbard is characterized by low-growing plants, with both shrubs and herbs reaching only a few centimeters (photo Jan Kavan); (**B**) The river floodplain in Kamchatka is dominated by gigantic herbs, *Petasites amplus* (photo by Kirill Korznikov) achieves its size through its leaves; (**C**) Although the oak is not the tallest tree in the world, it clearly shows that trees achieve their height by growing perennial stems

The weight of a seed pertains to an organ that we address here only marginally, yet it is worth mentioning why it is interesting. The weight of a single seed is relatively easy to determine, unlike the number of seeds produced by an individual during a single reproductive event. Plants can bloom gradually and shed seeds, or they can be enormous, making it difficult to count the seeds. Sometimes seeds are hidden in fruits, and to count them, the seeds must be extracted from the fruits. Fortunately, these two variables, the weight of a single seed and the number of seeds produced by a plant in one reproductive event, are correlated, so from the weight of one seed, we can infer the total seed production. The correlation is negative, meaning plants either invest in a large number of small seeds or a small number of large seeds. Small seeds are better dispersed in space and time because they are more easily spread far from the parental plant, and they survive better in the seed bank. However, the reserves for the growing seedling are small, and the plant struggles to establish itself in the shade. On the other hand, large seeds are less numerous, harder to disperse in time and space, but they provide good support for the plant in early growth and enable survival in shade or when injured.

2.2.3 Where Are We Now?

For the purpose of measuring functional traits, international teams have been established over the past 20 years, filling databases with measurements (Kattge et al. 2020), writing guidelines on how to correctly measure traits in different species (Pérez-Harguindeguy et al. 2013), and analyzing the first obtained data sets. A new discipline has emerged that tests the relationship between traits and functions: it uses easily measurable traits to predict the behavior of species in communities affected by various biotic and abiotic influences, to describe ecosystem functions such as vegetation flammability or the ability of plant litter to decompose quickly, to test the importance of species richness and functional richness for community functioning (for example, for biomass production or response to disturbance), to explore how environmental filters (such as high soil moisture) influence which species with which traits grow in the community; to determine whether there are general principles in the formation of plant communities. Researchers have measured and continue to measure various species traits, not just those proposed by the northern or southern school, and gradually they managed to gather data for a large set of species from around the world and across biomes. This enormous dataset, created during work on individual small studies and projects, was compiled into several databases, and truly global analyses were conducted with it.

What did the first major analysis of plant traits at the level of individual species based on worldwide data bring? (Díaz et al. 2016) Did it confirm or refute our notion that plants can be categorized into three strategies, as proposed by Philip Grime? The analysis showed us that plants are arranged according to two groups of traits. The first trait gradient is from small plants with low stem density and small seeds to large plants with large seeds and high stem density (along the x-axis in

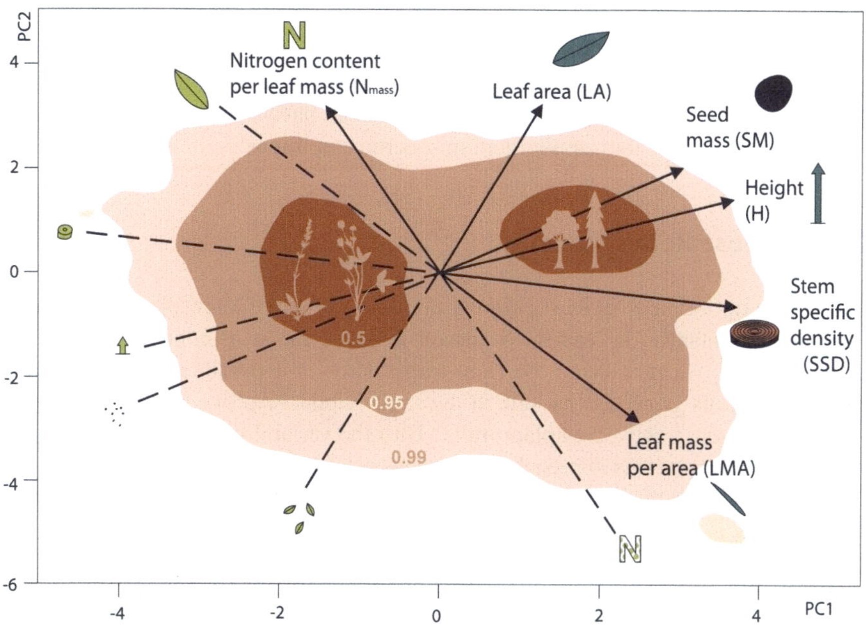

Fig. 2.15 The relationship between the six most measured plant traits according to their occurrence in plants across all biomes and continents. Areas shaded in various shades of pink indicate where the studied species concentrate in the space defined by six functional traits; one maximum corresponds to herbaceous species (left), which have small height, leaves, and seeds, as well as herbaceous stems, the other maximum corresponds to woody species (right) characterized by large height, seeds, and stem density. Arrows indicating traits: the solid part of the arrow indicates an increasing trait value, the dashed part of the arrow—a decreasing trait value, the sharper the angle between two arrows, the more correlated the occurrence of traits in species (example: the greater the height of the plant, the larger the seeds). (Redrawn according to Díaz et al. 2016)

Fig. 2.15). The second trait gradient is from plants with small leaves poor in nitrogen and with high biomass per leaf area, to plants with the opposite traits (y-axis in Fig. 2.15). The first gradient divides plants into herbs (on the left in Fig. 2.15) and woody plants (on the right), while the second gradient divides them into plants with conservative, resource-saving traits (at the bottom in Fig. 2.15), and resource-wasting plants (at the top). It seems that when looking at the global flora, we do not see three Grime categories, but rather four: fast and slow-growing herbs and fast and slow-growing woody plants. It appears as if the triangle should indeed be a square, the question is whether those small plants with small leaves truly inhabit environments where disturbance combines with nutrient scarcity.

The first analyses of plant traits at the community level (i.e., species that occur together), and the environmental conditions in which the community (Bruelheide et al. 2018) is found, confirmed that plant traits change along environmental gradients. However, the trait gradients found at the individual level, described by plant size and leaf quality (see Fig. 2.15), were only weakly correlated with climatic and soil properties on a global scale. Instead, it seems that the combination of species traits within a community is preferentially determined by local conditions, such as competitive relationships within the community, local soil properties, or disturbance regimes. It thus appears that plants do indeed have strategies, but we do not fully understand how they apply them in specific conditions.

2.2.4 Those Roots Again!

Now is the right time to at least partially question the use of a few plant traits, measured predominantly on leaves, to answer a whole range of complex ecological questions! Are we measuring the right traits? How is it possible that plants with contrasting traits grow side by side in the same habitat? Are we measuring habitat conditions at least, as well as plant traits? Isn't the variability in trait values within species so large that this approach makes no sense? And why does functional ecology not work with the functions of all basic organs? Why does it focus mainly on leaves? Why is the height of the stem and its function in light competition important, but we are not interested in its connective or regenerative function at all? And what about roots?

After the publication of the first global functional ecological analyses, rhizologists, scientists researching roots, as well as scientists studying other belowground organs, pointed out that the assumption that belowground and aboveground functions and plant traits are closely correlated may not hold. Having thin, large, short-lived leaves that have rapid photosynthesis, are palatable to herbivores, are easily decomposable, and are nutrient-demanding does not necessarily mean that your roots are set for maximum performance, meaning that they are thin, highly branched, short-lived, and capable of providing the plant with many nutrients, as was assumed. These doubts have led to a renaissance of interest in root traits, which we have so far overlooked due to their belowground location and the difficulty of measuring anything on them.

What do we know about root properties? The absorption of water and dissolved mineral substances occurs mainly through fine roots, although thick roots and especially mycorrhizal fungi also play a role. The ratio between the length and weight of fine roots (specific root length), branching architecture, root diameter, colonization by mycorrhizal fungi, nitrogen content, and persistence are parameters by which function can be estimated, similar to leaves on an economic gradient (Freschet et al. 2021) (Table 2.4). Rhizologists assembled existing data on fine root traits into a database and performed the first analyses; they found out that, similarly as with leaf traits, the fine root traits are intercorrelated (Bergmann et al. 2020). This time, however, it is not the trait space of plant size and plant economy, but trait space of cooperation and economy. The main axis of fine root strategies is determined by a gradient of dependency of resource acquisition on mycorrhizal symbionts versus self-service by plants investing in highly branched and thin roots without fungal

Table 2.4 Challenges plants face belowground and the traits they use to address them

Challenge		Complex trait	Simple trait (examples)
Space occupation	Large scale	Volume of root system in the soil	Maximum rooting depth, lateral extent of rhizomes and roots, proportion of root biomass on the whole plant biomass
	Small scale	Density of root system in the soil	Specific root length, root density, root branching, mycorrhiza
Acquisition of water and nutrients	Directly	Effectivity in nutrient acquisition	Content of N, P in fine roots, ratio between N and C in fine roots
	Using symbiosis	Effectivity in nutrient acquisition	Mycorrhiza, symbiosis with nitrogen fixing bacteria
Persistence and resprouting	After seasonal dormancy or damage	Storage carbohydrates and bud bank	Size of storage organ, number of buds
Resistance to pathogens, herbivores and parasites		Resistance	Mycorrhiza, anatomy
Anchorage in soil		Resistance to uprooting, protection of soil from erosion and landslides	Root diameter, root dry matter content, lignin content, depth of rooting, root branching

According to Freschet et al. (2021). *N* nitrogen, *P* phosphorus, *C* carbon

symbionts. Another axis of specialization that is independent from the first one, similarly as in leaves, is from effective, short lived, and rich in nitrogen fine roots, to less effective, poor in nitrogen, and long-lived fine roots.

The first analyses of root properties along environmental gradients were disappointing. For example, along an altitude gradient from forest to alpine grassland in the French Alps, although leaf traits varied as expected between individuals of the same species and between species, root traits did not respond at all (Weemstra et al. 2021). What was it caused by? There can be several reasons. Roots have a much more intense relationship with the environment in which they live, and this can change on a scale of millimeters independently of above-ground parameters such as air temperature, wind speed, air pressure, or radiation. Fine roots, unlike leaves, are responsible for finding and acquiring multiple resources that have different availabilities and distribution in the soil (Weemstra et al. 2016). And importantly, choosing a root to measure is not as simple as choosing a leaf to measure. While for leaves we can be careful to measure sun-drenched leaves that show no signs of biting and are well developed but not too old, for roots, we usually settle for whatever tender roots we can get from the soil, thus increasing the variability in measured values. So, there is still a lot of work ahead of rhizologists if they want to understand the functional properties of fine root traits.

2.3 Specialized Belowground Organs

2.3.1 Rhizomes, Bulbs, and Tubers

Let us recall that from Theophrastus's work, we know that there are not only roots belowground, and even if we call them roots, we must admit that they may have a different function. Belowground organs that are not fine roots with an absorptive function are either storage, regenerative, or clonal growth organs (or combine multiple functions). Plants do not only compete by their height in the vertical direction but also occupy space in the horizontal direction; they do not only reproduce by seeds but also by clonal growth; they do not only seek resources using roots or leaves but also using rhizomes or runners; they do not only regenerate after disturbance from the seed bank (i.e. viable seeds in the soil) but also from the bud bank (Reserve dormant buds on belowground plant organs) (Fig. 2.16). Organs ensuring these functions can be of stem, root, or leaf origin, and often involve a combination of different organs (Fig. 2.17 and Table 2.5). Unlike the worldwide consensus on how to define the basic organs of plants, these specialized organs often have names that are typical only for a given ecosystem, while elsewhere the organs may not occur or occur but are called differently. It is also a very diversified group of organs and functions, which hinders their understanding and also prevents the generalization of results obtained in one region (Klimešová et al. 2019).

For example, the rhizome of the nettle is a stem located belowground, serving the herb as a storage and regenerative organ because it is the only overwintering part,

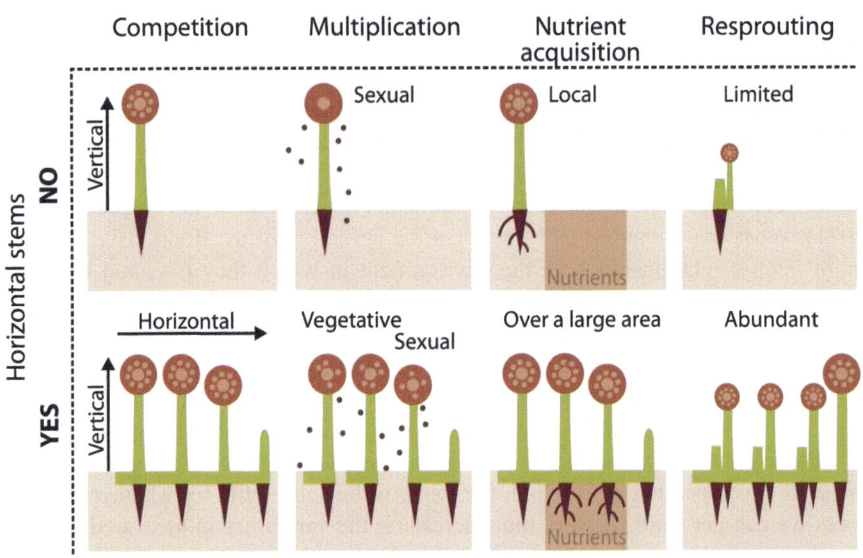

Fig. 2.16 Advantages of plants with horizontal stems (rhizomes, stolons) compared to plants without horizontal stems, in competition, reproduction, nutrient acquisition, and regeneration after disturbance. According to (Klimešová et al. 2018)

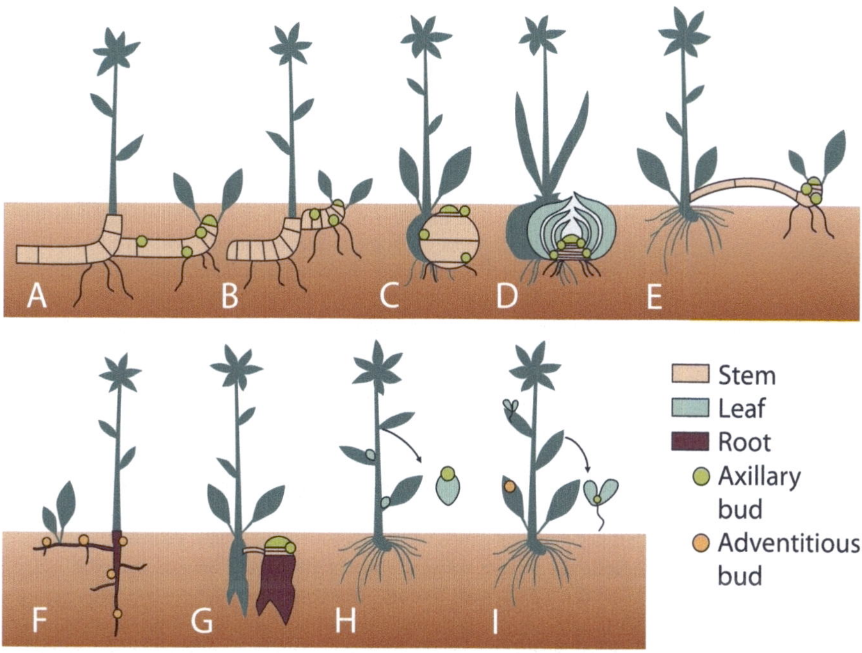

Fig. 2.17 Organs bearing a bud bank and serving to varying degrees for storage, regeneration, or clonal growth. (**A**) hypogeogenous rhizome, (**B**) epigeogenous rhizome, (**C**) stem tuber, (**D**) bulb, (**E**) stolon, (**F**) root with adventitious buds, (**G**) root tuber, (**H**) bulbil in the leaf axil on the above-ground part of the plant, (**I**) adventitious buds on leaves. For a detailed description, see Table 2.5. According to (Klimešová et al. 2019)

and above-ground stems with leaves regenerate from it. The plant can also use the rhizome to regenerate after disturbance by grazing or mowing. As the rhizome grows and branches on one side and simultaneously dies on the other, it gives rise to physically separate plants and also serves for clonal growth. A similar organ, which mainly differs by being woody, is called a xylopodium in Brazilian tropical grasslands and does not serve for overwintering but for regeneration after a dry period or fire. The stolon in the strawberry, a stem growing above ground and producing offspring plants, only serves for clonal growth because it does not survive the winter, and the offspring plants become independent from the rest of the clone. Unlike the rhizome, the stolon usually does not have a storage or regenerative function because it contains relatively small reserves of carbohydrates and buds. Tropical shrubs may have a similar clonal growth organ (Fig. 3.11, I), which we would hardly call a stolon because it is woody and perennial. Some root tubers, for example in *Hylotelephium telephium*, serve only as storage organs because they do not have buds from which they would regenerate or else they would use them to produce new parts of the clone. Other root tubers have buds because they connect to a stem that bears buds (as in many of terestrial orchids), or they are capable of creating adventitious buds, like sweet potato or *Filipendula vulgaris*, if they are disturbed (Bartušková

Table 2.5 Description of organs bearing a bud bank and serving to varying degrees for storage, regeneration, or clonal growth

Organ	Description
A Hypogeogenous rhizome	A belowground organ of stem origin. This rhizome forms belowground; grows from a belowground bud, continues to grow horizontally at a species-specific depth, and after some time turns upwards and forms a leafy shoot above-ground. The horizontal belowground part of the rhizome bears leaves transformed into scales and adventitious roots; stem internodes are mostly long. The hypogeogenous rhizome usually allows the plant to spread over a large distance (up to tens of centimeters per year) and its persistence varies between species. Examples: *Urtica dioica*, *Agropyron repens*, *Aegopodium podagraria*
B Epigeogenous rhizome	A belowground organ of stem origin. A rhizome formed above-ground and only later covered by soil and litter or pulled belowground by contractile roots; in the ground, it can be positioned horizontally, obliquely, or even vertically; nodes bear green leaves, sometimes alternating with leaves transformed into scales; it bears adventitious roots; stem internodes are mostly short. The epigeogenous rhizome allows the plant to spread mostly only over a short distance, and its persistence varies between species. Examples: *Succisa pratensis*, *Geranium pratense*, *Cirsium oleraceum*
C Stem tuber	A belowground organ of stem origin. Usually a short-lived organ, formed by one or several thickened stem internodes; offspring tubers arise either sessile on the parental tuber or at the end of a belowground rhizome. The connection between offspring tubers is usually short-lived and tubers can vary in size, and subsequently in their ability to regenerate. Examples: *Corydalis solida*, *Colchicum autumnale*, *Trientalis europaea*
D Bulb	A belowground organ of stem (buds) and leaf (storage) origin. The bulb usually has a shortened stem that bears storage scales protecting the growth apex. The bulb can be annual or perennial; it can produce equally sized offspring bulbs (usually in small numbers) or small reproductive bulbils (which can be numerous). Examples: *Galanthus nivalis*, *Allium ursinum*, *Lilium bulbiferum*
E Stolon	An above-ground organ of stem origin. It is mostly a horizontally growing stem with long internodes rooting at nodes or at the end of the annual growth. It usually spreads over a large distance (tens of centimeters) and produces one or several offspring, surviving mostly only one season. Examples: *Ranunculus repens*, *Fragaria vesca*, *Potentilla anserina*
F Root with adventitious buds	A belowground organ of root origin. These roots (main root, hypocotyl, and/or adventitious roots) capable of forming adventitious buds spontaneously or after injury; if they are horizontally growing roots, there can be significant lateral spread of the clone. Examples: *Sonchus arvensis*, *Cirsium arvense*, *Robinia pseudoacacia*
G Root tuber	A belowground organ of stem (buds) and root (storage) origin. It is usually a short-lived organ formed by a shortened stem and a thickened adventitious root. Offspring tubers arise sessile to the parental tuber and are replaced annually. Root tubers that do not bear buds are only storage organs and cannot serve for regeneration or clonal growth. Examples: *Ficaria verna*, *Dactylorhiza majalis*, *Anthriscus sylvestris*

(continued)

Table 2.5 (continued)

Organ	Description
H Bulbil in the leaf axil on the above-ground part of the plant, turion in aquatic plant	A small above-ground reproductive bulbil similar to a bulb, root tuber, or stem tuber, forming in the axil of leaves of the above-ground part of the plant or in the inflorescence (called pseudovivipary). Its connection with the parent plant is very short-lived, therefore a small offspring plant similar to a seedling grows from it. Plants with this type of clonal growth have the potential for large-scale spread over long distances. In aquatic plants, compact buds, turions, serve for overwintering, and easily spread over long distances by water. Examples of terrestrial plants: *Lilium bulbiferum*, *Dentaria bulbifera*, *Saxifraga cernua*; examples of aquatic plants: *Potamogeton crispus*, *Lemna minor*, *Aldrovanda vesiculosa*
I Adventitious buds on leaves	Above-ground organ of leaf origin. A leaf capable of producing adventitious buds spontaneously or after disturbance. The significance of this type of clonal growth for temperate herbs is unknown. Examples: *Cardamine pratensis*, *Drosera rotundifolia*
J Perennial main root	The plant is not capable of forming adventitious roots and the main root persists throughout its life, although in old individuals it may disintegrate from the center. Buds are on the perennial bases of stems form the so-called root crown, and carbohydrate reserves are in the root. Examples: *Taraxacum officinale*, *Salvia pratensis*, *Verbascum nigrum*
K Perennial hypocotyl tuber	The plant forms a tuberous organ for storage from the hypocotyl, which persists for the entire life of the plant and from which adventitious roots grow. The reserve of buds is on the perennial stem bases. Examples: *Cyclamen purpurascens*, *Corydalis cava*, *Rhodiola rosea*

A–K denote schematic pictures of organs in Fig. 2.20. All organs except the perennial main root and hypocotyl tuber (J, K) are clonal organs; the stolon (E) is an above-ground organ that serves for clonal growth but has little significance for the regeneration of an individual plant after disturbance, although it enables fast occupation of empty space that is created by disturbance.

et al. 2017). Root tubers are often found in other biomes among herbs, shrubs, and trees. If a woody plant has a thickened root with a reserve of buds, this organ, which does not occur in temperate flora, is called a lignotuber and is considered an adaptation for surviving fire.

How can we create a system that would allow us to compare the function of belowground organs that are not fine roots across different species? Their structure usually involves multiple basic organs, so it is good to distinguish which organ bears the buds, which organ stores carbohydrate reserves, and which connects the individual parts of the clone if the plant is clonal. Other characteristics are also important, such as whether the derived organ arises above- or belowground, where its final position relative to the soil surface is, how long it persists, or whether it becomes woody. The following classification example is valid for temperate climate biomes (deciduous forests, boreal forests, steppes, and arctic tundra); if it were to be used globally, it would require some adjustments.

Since the most crucial aspect for a plant's life is the ability to bear buds, we will take the organ bearing buds as the basis for classification and divide all belowground organs into those that have dormant buds (i.e., a bank or reserve of buds, see the chapter on it) on stem organs, root organs, or leaf organs. We will also consider where the carbohydrate reserves are stored, which serve the plant in case it loses its above-ground green parts and cannot photosynthesize. Carbohydrate reserves can be directly in the stem that bears the buds, but also in other organs. Carbohydrate and bud reserves are usually linked in plants because when the plant fragments, it needs both for regeneration, and it makes no sense to store buds separately from energy and building blocks (i.e. carbohydrates). The same stem usually bears buds and is a storage organ, additionally capable of producing adventitious roots and forming connections between parts of the clone. Such a stem is called a rhizome, and it is the most common organ of clonal growth in temperate flora.

Rhizomes have various shapes and can differ in the intensity of clonal growth, storage function, and regeneration ability (see the chapter on clonal traits). The most important feature for distinguishing basic types of rhizomes is how they grow: they are formed either on the soil surface and only later get belowground being pulled by contractile roots, these are called epigeogenous rhizomes (for example, in *Succisa pratensis*), or rhizomes form belowground, and these are called hypogeogenous rhizomes (for example, in *Urtica dioica*). These two types of rhizomes were distinguished by Russian botanists, the Serebriakovs (Serebriakov and Serebriakova 1965), and they fundamentally differ in their function, even though they look almost the same at first glance. We can imagine epigeogenous rhizomes as palm trunks pulled belowground. The herb that produces them has a leaf rosette on the stem, like a palm, but the stem does not grow upwards and does not form a trunk; instead, it is pulled into the soil by roots so that the growth tip is at the soil surface, where it is protected from frost. Hypogeogenous rhizomes grow from belowground buds and grow horizontally below the soil surface, producing only scale leaves until they reach their typical length. Then their growth tip turns upwards and grows aboveground as a leafy shoot.

Belowground organs of stem origin are the most common in nature and have the most modifications, mainly because they bear the buds because storage carbohydrates can be in other organs (bulbs have them in leaves), because they are growing aboveground (stolon), or they are very short and thick (tuber). Organs based on adventitious branching from roots are rarer and do not have as many modifications, and those based on leaves are rather a rarity in our flora (Table 2.5 and Fig. 2.17).

2.3.2 What Do We Know About Evolution?

The oldest terrestrial spore-bearing plants were composed of a stem that grew upwards, branched dichotomously, but then gradually laid down on the ground. The stems pressed to the ground took root and died from the oldest parts, leading to the disintegration of the clone. Horizontal stems could also be buried by the substrate, for example, sediment in river floodplains (Xue et al. 2016). Later, tree forms of spore-bearing plants and gymnosperm trees evolved, which do not have any specialized belowground organs and are mostly not clonal. Exceptions are shrubs with prostrate branches that can become independent over time (for example, dwarf pine), groups of trees formed by rooting lower branches in various species of spruce at the alpine tree line, or araucarias sprouting from adventitious buds on the roots.

In angiosperms, especially in herbs, specialized belowground organs are common. For example, for monocotyledonous plants with stem tubers, bulbs, and rhizomes, a great conservatism in evolution was found, where a given organ (bulb or rhizome) is typical for an entire large taxonomic group (Howard et al. 2019). Conversely, in the analysis of clonal organs of the entire Central European flora, it was found that the evolution of clonality and a given clonal organ is flexible, and transitions are common, for example, between plants with epigeogenous and hypogeogenous rhizomes. Other transitions, such as the transition between stem-based clonality (various types of rhizomes, tubers, and bulbs) and root-based clonality (adventitious sprouting from roots), are rare (Herben and Klimešová 2020) (Fig. 2.18). It thus appears that acquiring a derived belowground organ and becoming a clonal plant have been relatively easy, and there are two independent paths to

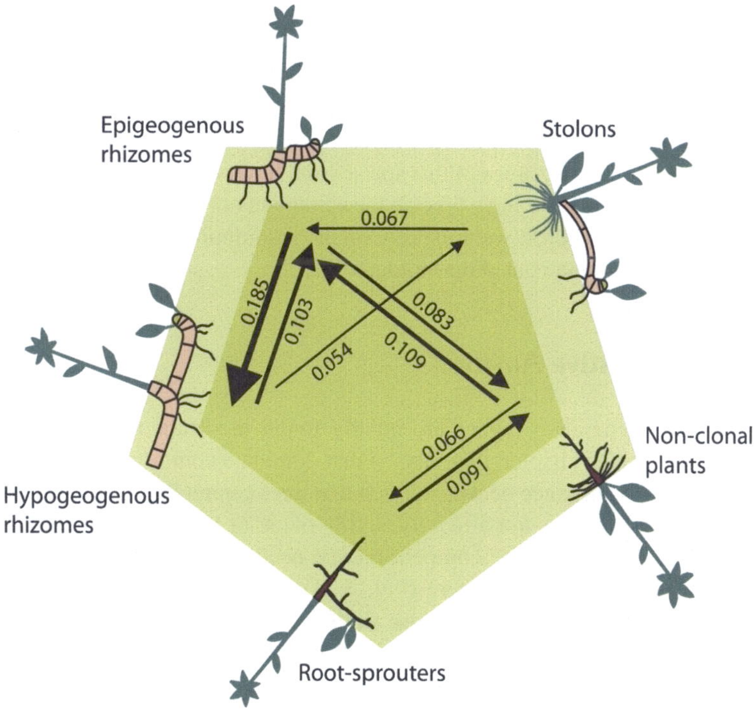

Fig. 2.18 Probability of evolutionary transitions between different types of clonal growth in the evolution of the Central European flora. The most common transition is between epigeogenous and hypogeogenous rhizomes, and other fairly common transitions are from non-clonal plants to plants with epigeogenous rhizomes or to root sprouting. Stem-based clonality (rhizomes, stolons) is independent of root-based clonality (plants sprouting from roots). According to Herben and Klimešová (2020)

it, through the stem and through the root. Whether these conclusions will be confirmed for other floras, we do not know; we would need to know the organs of clonal growth in more plants from different floras and ecosystems.

2.3.3 Storage Function

We refer to storage organs as those where the plant stores assimilates for periods when it will not be able to photosynthesize, for example, due to disturbance of above-ground biomass or winter dormancy. Assimilates are various types of carbohydrates, such as starch, inulin, sucrose, fructose, and glucose; for simplicity, we will simply call them carbohydrates here. How do we recognize which plant organ is a storage organ when carbohydrates are found in all plant organs? Interestingly, we distinguish storage organs quite intuitively by the fact that they are thicker than organs of the same type usually are. For example, the roots of *Filipendula vulgaris* have thickened sections, root tubers, which are more than 10 times the diameter of the roots on which they are found and represent a larger biomass than the rhizome of the given species (Fig. 2.19). These root tubers also contain a higher concentration of storage carbohydrates than rhizomes and a greater total reserve when considering the size of individual organs. However, this may not always be the case. Another species, *Cirsium pannonicum,* has normal non-thickened roots, but they contain a higher concentration of carbohydrates than rhizomes. The total reserves stored in the roots and rhizomes of this species are, however, the same (Janeček and Klimešová 2014). Since we have precise measurements of carbohydrate concentrations for only a few species, we must rely on the assumption that the most voluminous belowground organ is also the storage organ.

2.3.4 Regenerative Function

Belowground regenerative organs are mainly found in shrubs and herbs, that is, growth forms (see the chapter about them) that dominate communities where some type of repeated disturbance occurs so often that trees cannot live there. At the same time, the disturbance does not affect the soil horizon, or there are few nutrients, so annuals, which cannot tolerate competition and need sufficient resources for their rapid growth, do not establish there (Fig. 3.24). Thanks to belowground organs, these perennial herbs and shrubs can lose all their above-ground biomass and regenerate from belowground organs (Fig. 2.19). The belowground organs provide them with reserve buds, from which new shoots can grow, and reserve carbohydrates, which supply the building blocks for new growth (see also the Sect. 3.4).

Comparative studies proving that morphological types of belowground organs specialize in a certain disturbance regime are still rare. The results of the analysis of Czech flora show that the most sensitive to disturbance are runners, then

Fig. 2.19 Belowground organs of *Filipendula vulgaris*. The largest root tuber (MT) was separated from the parent plant by some disturbance (thick arrow) and a new plant grew from it, forming an epigeogenous rhizome (Rh). Adventitious roots with tuberous thickenings (RT) grow from the rhizome. MT parental tuber, Rh rhizome, AdR adventitious roots, RT root tubers. (Photo Š. Janeček) is the author of photograph

epigeogenous rhizomes, hypogeogenous rhizomes, perennial main roots, and the best-performing plants are those sprouting from roots, which are found in really frequently and strongly disturbed places, such as arable land (Klimešová and Martínková 2022) (Fig. 2.20). For Central Europe, soil cultivation is the most widespread large-scale strong and frequent disturbance (it is far to the right in Fig. 3.24). In agriculture, the disturbance regime has significantly changed over the last century, as it transitioned from manual mechanical weeding, through machine mechanical weeding, to herbicides.

In each stage, weeds with different properties dominated because the method of weeding selected specific weed strategies in Central Europe. The greatest diversity of weeds was at the beginning of this period when fieldwork was mostly manual, and in the weed community, annuals, biennials, and perennials usually coexisted because the mechanical weeding was not perfect. Perennials had a whole range of strategies for vegetative reproduction and regeneration. Short-lived species capable of sprouting from the hypocotyl and roots, such as *Kickxia spuria* and *Anagallis arvensis*, were typical for root crops, where disturbance occurs repeatedly during the season. In vineyards, there was a specific weed community in which annuals and bulbous plants like *Gagea pratensis*, *Ornithogalum umbellatum*, or *Tulipa sylvestris* were present.

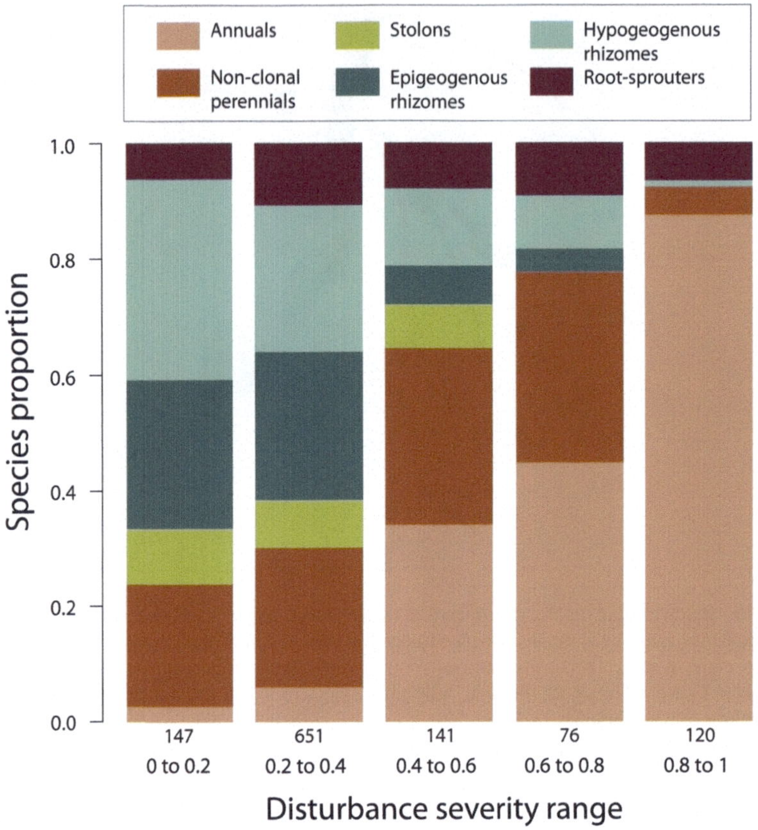

Fig. 2.20 Proportional representation of herbaceous species with different regenerative organs along the disturbance severity gradient in Central Europe. While the mildest disturbance (0–0.2) corresponds to forests, the most severe disturbance (0.8–1) corresponds to arable land. While the proportion of annuals (regenerating from seeds) increases with disturbance severity, the share of all other categories, except for adventitious root sprouting, decreases. For a description of regenerative organs, see Table 2.5. Numbers under the columns indicate the number of species occurring in individual levels of disturbance severity. Redrawn according to (Klimešová and Martínková 2022)

With the development of mechanization, the number of species regenerating from belowground regenerative organs began to decrease because they were destroyed by deep plowing (Lososová et al. 2004). Perennials that survived the overturning of the soil profile to a depth of 20 centimeters were mainly species sprouting from roots stored deep in the soil, such as *Cirsium arvense* or *Sonchus arvensis*.

With the advent of herbicide use, the composition of the weed community changed even more in favor of annual weeds (Johnson et al. 2009), because perennials do not have much chance to avoid the effect of the poison, and due to the slow turnover of generations, it is difficult for them to develop resistance to chemical agents. However, among the reported resistant weeds, perennials with belowground regenerative organs have recently begun to appear (Holt et al. 2013). It holds true here that if we use one method of weed control, we will select types adapted precisely to that method of weed control, whether by timing the life cycle, storing a bud bank, or the ability to create buds on otherwise budless parts of the plant body.

Purely mechanical soil processing in agriculture still exists, especially in organic farming that deliberately does not use herbicides, or where there is a lack of resources for herbicides. Plowing leads to the fragmentation of belowground weed organs, against which this method of management is targeted. Although annual weed species, which regenerate after disturbance from seeds and are capable of producing new seeds before the next disturbance (as predicted by the B&S model in Fig. 3.24), predominate on arable land, there is also a group of perennial plants. These plants are characterized by their organs carrying a short-lived bud bank and spreading over long distances (Herben et al. 2018). They are mostly horizontal roots carrying adventitious buds or long hypogeogenous rhizomes. These long structures are, of course, easily fragmented by plowing, and each fragment can give rise to a new plant. Plowing weakens the perennial weed but simultaneously multiplies and spreads it across the entire field or to other plots.

Less commonly than rhizomes or horizontal roots with adventitious buds, perennial weeds have reproductive belowground vegetative diaspores, which easily detach from the parent plant, such as tubers or bulbs. In the Central European weed flora, for example, these are the already mentioned rare vineyard weeds. In the Mediterranean, the invasive species, *Oxalis pes-caprae*, originally from South Africa, is very successful, spreading on arable land even though it is a pentaploid incapable of producing seeds (Papini et al. 2017). It manages to reproduce using bulbs (one plant produces up to 25 bulbs per year), which are unintentionally transported from field to field by agricultural machinery. Because it is a new plant in the area and very conspicuous with its yellow flowers, it is easy to track the significance of regenerative organs and human activity for the spread of perennial weeds in the landscape by naked eye.

2.3.5 Clonal Growth

We refer to clonal growth organs as those capable of creating independent clonal offspring—ramets—through their growth. Clonal growth organs are not only belowground (see examples above). While the most common belowground clonal organ in our flora is the rhizome, the most common above-ground clonal organ is the stolon. Various types of reproductive bodies in the axils of leaves of above-ground shoots or directly in the inflorescence also occur (see Fig. 2.17 and Table 2.5).

A great diversity of clonal growth organs can be found in aquatic plants, which often combine several options: rhizomes, stolons, tubers, and turions (Klimeš and Klimešová 1999). This is because the aquatic environment is ideal for clonal growth but may be risky for flowering and sexual reproduction. Many aquatic plants do not

Fig. 2.21 Turion, an overwintering organ of aquatic plants, which also serves for clonal reproduction. The turion has a structure similar to a bulb: it has a shortened stem, and storage leaves clustered into a compact formation. You cannot find turions during the growing season; they form in the autumn. The development cycle and position in the water column (blue) of a plant overwintering with turions is indicated by the diagram at the top of the figure. T_1—non-rooting aquatic plant during the growing season floating close to the water surface; T_2—freshly formed turion floating close to the water surface; T_3—overwintering turion being dormant at the bottom of the reservoir; T_4—germinating turion in spring, when it floats back to the surface. (**A**) Shoot apices of *Utricularia macrorhiza* with freshly formed turions; (**B**) overwintering turions separated from the parent plant of *Utricularia vulgaris*; (**C**) winter turions of *Aldrovanda vesiculosa*; (**D**) germinating turions of *Caldesia parnassifolia*; (**E**) germinating turions of *Utricularia vulgaris*; (**F**) germinating turions of *Hydrocharis morsus-ranae*. (All photos by Lubomír Adamec)

root in bottom and every fragment of the stem immediately becomes an independent plant. A specialty of aquatic plants in temperate climates is the production of turions, which are essentially buds with clustered leaves (similar to bulbs, but green). The plant produces turions in the fall, and they sink to the bottom of the water, where they survive the winter, rising to the surface in spring and continuing to grow (Fig. 2.21). Because one parent plant usually produces multiple overwintering turions, these bodies serve not only to survive unfavorable periods but also for clonal multiplication.

Conversely, dry environments are not favorable for clonal plants, probably because it may be difficult to produce new roots on new rhizome increments. Belowground clonal organs also do not thrive in rocky and mobile soil, for example, in the Arctic, where permafrost and frost movements tear roots and rhizomes. However, it is advantageous here to produce clonal offspring on the above-ground parts of the plant, and we find a relatively large number of plants with bulbils and pseudo-viviparity (Klimešová et al. 2012), for example, in various species of saxifrages (*Saxifraga cernua*), grasses (*Poa alpina*), and in *Bistorta vivipara* (see also pseudoviviparity in Table 2.5).

The complete overview of the clonal growth organs in the species of a flora is available only for Central Europe in the CLO-PLA database (Klimešová et al. 2017) (Box 11). This database stores data on the clonality (and non-clonality) of plants for the entire flora, listing morphology of clonal growth organs as well as clonal traits (see Sect. 3.5).

References

Arber A (1920) Water plants. Cambridge University Press, A study of aquatic Angiosperms

Arber A (1934) The Gramineae. A study of cereal, bamboo, and grass. Cambridge University Press

Armani M, Charles-Dominique T, Barton KE, Tomlinson KW (2019) Developmental constraints and resource environment shape early emergence and investment in spines in saplings. Ann Bot 124:1133–1142

Bartušková A, Malíková L, Klimešová J (2017) Checklist of root-sprouters in the Czech flora: mapping the gaps in our knowledge. Folia Geobot 52:337–343

Bartušková A, Filartiga AL, Herben T, Qian J, Klimešová J (2021) Comparative analysis of root sprouting in temperate herbs: anatomical correlates and environmental predictors. Ann Bot 127:931–941

Bergmann J, Weigelt A, van Der Plas F, Laughlin D, Kuyper TW et al (2020) The fungal collaboration gradient dominates the root economics space in plants. Sci Adv 6:eaba3756

Bruelheide H, Dengler J, Purschke O, Lenoir J, Jiménez-Alfaro B et al (2018) Global trait–environment relationships of plant communities. Nat Ecol Evol 2:1906–1917

Classen-Bockhoff R (2016) The shoot concept of the flower: still up to date? Flora 221:46–53

Díaz S, Kattge J, Cornelissen JHC, Wright IJ, Lavorel S, Dray S, Reu B et al (2016) The global spectrum of plant form and function. Nature 529:167–171

Freschet GT, Roumet C, Comas LH, Weemstra M, Bengough AG et al (2021) Root traits as drivers of plant and ecosystem functioning: current understanding, pitfalls and future research needs. New Phytol 232:1123–1158

Goethe JW (1790) Versuch die Metamorphose der Pflanzen zu erklären. Ettingersche Buchhandlung, Gotha

Grime JP (1977) Evidence for the existence of three primary strategies in plants and its relevance to ecological and evolutionary theory. Am Nat 111:1169–1194

Grime JP, Hodgson JG, Hunt R (2007) Comparative plant ecology: a functional approach to common British species, 2nd edn. Castlepoint Press, Colvend

Herben T, Klimešová J (2020) Evolution of clonal growth forms in angiosperms. New Phytol 225:999–1010

Herben T, Klimešová J, Chytrý M (2018) Plant traits along productivity and disturbance gradients in a temperate flora. Funct Ecol 32:799–808

Holt JS, Welles SR, Silvera K, Heap IM, Heredia SM et al (2013) Taxonomic and life-history bias in herbicide resistant weeds: implications for deployment of resistant crops. PLoS One 8:e71916

Howard CC, Folk RA, Beaulieu JM, Cellinese N (2019) The monocotyledonous underground: global climatic and phylogenetic patterns of geophyte diversity. Am J Bot 106:1–14

Humboldt, A. (2011) Ideas for a physiognomy of plants. In: Otté E, Bohn H (trans) Views of nature: or contemplations on the sublime phenomena of creation, Cambridge library collection – Earth science. Cambridge University Press, pp 210–352

Janeček Š, Klimešová J (2014) Carbohydrate storage in meadow plants and its depletion after disturbance: do roots and stem-derived organs differ in their roles? Oecologia 175:51–61

Johnson WG, Davis VM, Kruger GR, Weller SC (2009) Influence of glyphosate-resistant cropping systems on weed species shifts and glyphosate-resistant weed populations. Eur J Agron 31:162–172

Kattge J, Bönisch G, Díaz S, Lavorel S, Prentice IC, Leadley P, Tautenhahn S et al (2020) TRY plant trait database—enhanced coverage and open access. Glob Chang Biol 26:119–188

Kenrick P, Strullu-Derrien C (2014) The origin and early evolution of roots. Plant Physiol 166:570–580

Kirchoff BK (2001) From Agnes Arber to new explanatory models for vascular plant development – Preface. Ann Bot 88:1103–1104

Klimeš L, Klimešová J (1999) CLO-PLA2 - a database of clonal plants in Central Europe. Plant Ecol 141:9–19

Klimešová J, Martínková J (2022) Chapter 11: Clonal growth, resprouting, and vegetative propagation of weeds. In: Upadhyaya MK, Clements DR, Shrestha A (eds) Persistence strategies of weeds. Wiley Blackwell, pp 200–218

Klimešová J, Doležal J, Prach K, Košnar J (2012) Clonal growth forms in Arctic plants and their habitat preferences: a study from Petuniabukta, Spitsbergen. Polish Polar Res 33:421–442

Klimešová J, Danihelka J, Chrtek J, de Bello F, Herben T (2017) CLO-PLA: a database of clonal and bud bank traits of Central European flora. Ecology 98:1179

Klimešová J, Martínková J, Herben T (2018) Horizontal growth: an overlooked dimension in plant trait space. Perspect Ecol Evol Syst 32:18–21

Klimešová J, Martínková J, Pausas JG, de Moraes MG, Herben T et al (2019) Handbook of standardized protocols for collecting plant modularity traits. Perspect Ecol Evol Syst 40:125485

Lososová Z, Chytrý M, Címalová S, Kropáč Z, Otýpková Z et al (2004) Weed vegetation of arable land in Central Europe: gradients of diversity and species composition. J Veg Sci 15:415–422

Owen R (1848) On the archetype and homologies of the vertebrate skeleton. Richard and John R. Taylor, London

Papini A, Signorini MA, Foggi B, Della Giovampaola E, Ongaro L, Vivona L et al (2017) History vs. legend: retracing invasion and spread of Oxalis pes-caprae L. in Europe and the Mediterranean area. PLoS One 12:e0190237

Penzig O (1921–2) Pflanzen Teratologie, 2nd edn. Verlag von Gebrüder Borntraeger

Pérez-Harguindeguy N, Díaz S, Garnier E, Lavorel S, Poorter H, Jaureguiberry P, Bret-Harte MS et al (2013) New handbook for standardised measurement of plant functional traits worldwide. Aust J Bot 61:167–234

Serebriakov IG, Serebriakova TI (1965) On the two types of the rhizome formed by perennial plants. Biulletin Moskovskogo Obshchestva Ispitatelei Prirody, Biologia 70:67–81. (in Russian)

Tomescu AMF (2009) Megaphylls, microphylls and the evolution of leaf development. Trends Plant Sci 14:5–12

Troll W (1941–42) Vergleichende Morphologie der höheren Pflanzen. Verlag von Gebrüder Borntraeger, Berlin

Weemstra M, Mommer L, Visser EJW, van Ruijven J, Kuyper TW, Mohren GMJ, Sterck FJ (2016) Towards a multidimensional root trait framework: a tree root review. New Phytol 211:1159–1169

Weemstra M, Freschet GT, Stokes A, Roumet C (2021) Patterns in intraspecific variation in root traits are species-specific along an elevation gradient. Funct Ecol 35:342–356
Weiher E, van der Werf A, Thompson K, Roderick M, Garnier E, Eriksson O (1999) Challenging Theophrastus: a common core list of plant traits for functional ecology. J Veg Sci 10:609–620
Westoby M (1998) A leaf-height-seed (LHS) plant ecology strategy scheme. Plant Soil 199:213–227
Xue J, Deng Z, Huang P, Huang K, Benton MJ et al (2016) Belowground rhizomes in paleosols: the hidden half of an early Devonian vascular plant. PNAS 113:9451–9456

Plant Architecture 3

Plant bodies are composed of repeated motifs, modules, containing a stem internode and a node with a leaf and axillary bud. These modules together create shoots that are added and supplement older shoots, or that replace some older shoots during the lifespan of the plant. Where new shoots (i.e. branching) are formed, whether they are affected by flowering, how long shoots persist, and their relationship to roots determines plant architecture, which might be described as architectural models. Despite detailed classification of architectural models, we are only just beginning to use the principles of plant architecture for understanding plant functions, like clonal multiplication or bud bank building.

3.1 Growth Apex

3.1.1 Modular Structure

In the previous chapter, we mentioned that the plant body consists of leaves, stems, and roots, but we did not address their number and relative position. Now, however, we will try to look at the plant differently, noticing that the same motifs (modules) are repeated in its body, such as twigs in the crown of a tree. Let's look at how this modular structure of the plant arises, explain what its temporal and spatial arrangement (architecture) is, and try to answer the question of whether it has any ecological significance.

As always, we must start in ancient Greece, because Theophrastus considered the shoot (a twig produced by the tree in a given growing season) to be the basic plant organ. The shoot itself further consists of repeating structural units, just like all stems on the plant: an internode, a node, a leaf, and a bud in its axil. This smallest repeating module in the plant body can be an independent plant if it is capable of rooting, for example, a piece of creeping clover (Fig. 3.1). Sometimes this basic

J. Klimešová, *Plant Bodies*, https://doi.org/10.1007/978-3-031-92192-6_3

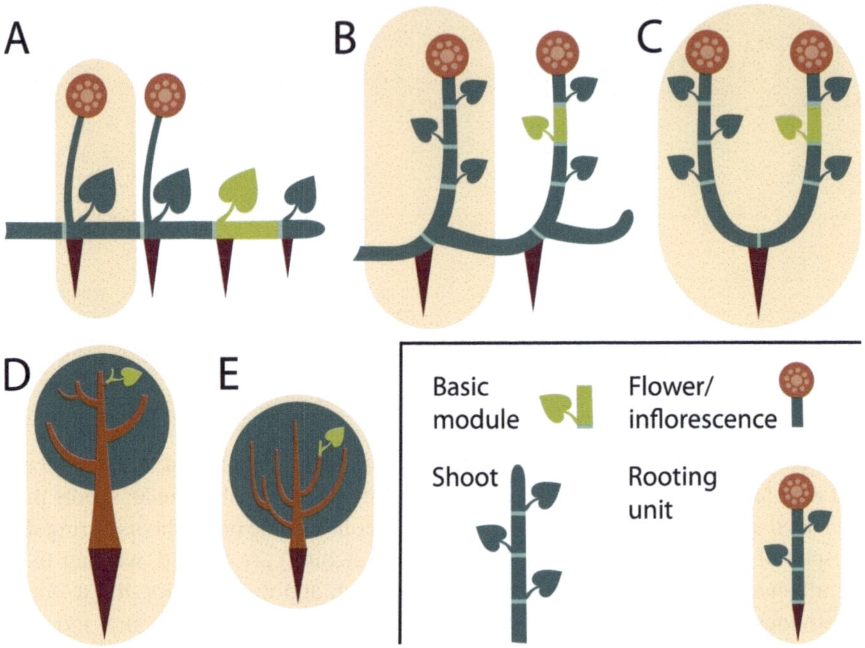

Fig. 3.1 Modular structure in various growth forms of plants. (**A**) The basic module is simultaneously a potentially independent rooting unit and thus a ramet of a clonal plant (*Trifolium repens*), (**B**) The basic module is part of the shoot, and the shoot is a potentially independent rooting unit and thus a ramet of a clonal plant (*Urtica dioica*), (**C**) The basic module and shoot are part of a rooting unit formed by several shoots connected with a single root system (*Dianthus carthusianorum*), (**D**, **E**) The basic module and shoot are part of a rooting unit formed by several shoots connected with a single trunk and root system (*Quercus* and *Coryllus*). (According to Klimešová et al. 2019)

module is part of a shoot (a leafy stem), which is an independently rooting unit. Furthermore, the basic module can be part of a shoot that, together with other shoots, shares one root, and then this entire set of shoots is a rooting unit.

The modular structure differs between growth forms because it is simpler in herbs than in woody plants. This difference points to the fact that plant organs, as we discussed in the previous chapter, are part of variously complex plant bodies. Should we ecologists concern ourselves with whether the leaf, whose function we are studying, is connected to the root only through a small piece of stem, as in creeping clover, or through a complicated system of hierarchically branching stems, as in a sequoia? In the following chapters, we will attempt to find an answer to this question. Because modularity and consequently architecture have been and are viewed from different angles, we will have to delve a bit into the past and also address topics that ecologists usually overlook.

3.1.2 Fountain of Youth

Theophrastus was the first to correctly describe how plants differ from animals. While plants lose and then renew some of their parts or simply add new parts each year, in animals, all parts of the body arise at the beginning of life and then are permanent. Plants, therefore, unlike animals, have an indefinite number of parts. How it is possible that plants keep growing was described by Caspar Wolf (Box 12), who discovered the growth apex. In his dissertation from 1759 "Theoria Generationis," he refuted the prevailing belief (which Theophrastus had already refuted for plants) that the entire organism (plant or animal) is already established in the embryo and only enlarges. So how does the plant grow? It is enabled by the growth apex (apical meristem), which contains dividing tissue that produces new stem segments with leaves and buds (potential future growth apices) or new roots.

A seed plant has two growth apices after germination (Fig. 2.1), one, growing positively geotropically, produces the root, and the second, growing negatively geotropically, produces a stem (bipolar growth, see also further). The root growth apex is simpler, producing only the main axis of the root, whereas lateral roots grow later independently from it, from the depth of the root tissue. The stem growth apex produces a modular structure of the stem by gradually establishing leaf primordia on its surface, which develop into leaves. In the axil of each leaf, there is a piece of meristematic tissue, another potential growth apex—the axillary bud. This axillary bud either immediately grows into a lateral branch of the stem, transforms into a flower, or remains dormant and serves as a reserve for the plant in case of loss of the active growth apex. All dormant buds on the plant then form a bud bank, i.e., a reserve of meristematic areas from which new parts of the plant can be produced after disturbance or seasonal dormancy.

The growth apex of the stem is well-studied from a developmental perspective. A germinating plant has only one such apex and is very vulnerable. Its fate determines whether it will be able to complete its life and produce seeds. The growth apex is composed of dividing cells, which are maintained in an undifferentiated state through the interaction of hormonal regulation. The growth apex can either remain forever in a vegetative state (continuously producing leaves) or, due to external and internal signals, switch from vegetative growth to generative growth, where the growth apex differentiates into floral organs. The growth apex can also gradually diminish as it produces more and more leaf organs until it disappears, or it can perish.

3.1.3 Corner's Rule

The basic parameters of the growth apex, such as size, number of cells, and its overall shape, correlate with some other plant characteristics: cell size with genome size and seed weight, while the number of cells correlates with stem diameter, leaf size, and seed weight (Schnablová et al. 2017). The variability of sizes and quantities of organs is in a mutual relationship known as Corner's Rule (Hallé et al. 1978). A small growth apex produces a highly branched stem with a small diameter and a large number of small leaves, whereas a large growth apex produces a thick unbranched stem, few leaves, and few buds (Corner 1949). Practically, we can imagine it as a gradient from a birch with thin twigs, small leaves, and a highly branched crown, to a chestnut with thicker twigs, large leaves, and less frequent branching, then to a palm with a huge growth apex, large leaves, and almost no branching (Fig. 3.2).

The same dependence on the size of the growth apex by its products as we know in trees is observable in herbs as well. Here, however, the consequences of these organ correlations can be even more interesting because branching can mean clonal reproduction, for example, when it comes to the growth of a belowground stem, i.e., a rhizome. Herbs with large leaves will thus have a less branched rhizome and less clonal multiplication than herbs with small leaves. Herbs with large leaves will also necessarily have fewer axillary buds in the annual increment of the rhizome. And because fewer large herbs fit per unit area than small herbs, the total reserve (bank) of buds per unit area of a community is also lower in vegetation with large herbs than in vegetation with small herbs. This effect must be considered when evaluating the size of the bud bank for different vegetation types (see further). The effect of Corner's Rule on the bud bank is interpreted by some authors as if the main selection pressure was disturbance and the main trait was the bud bank. However, the same effect would also result from selection pressure favoring the largest/smallest branching or the largest/smallest leaves (Whitman and Aarssen 2010).

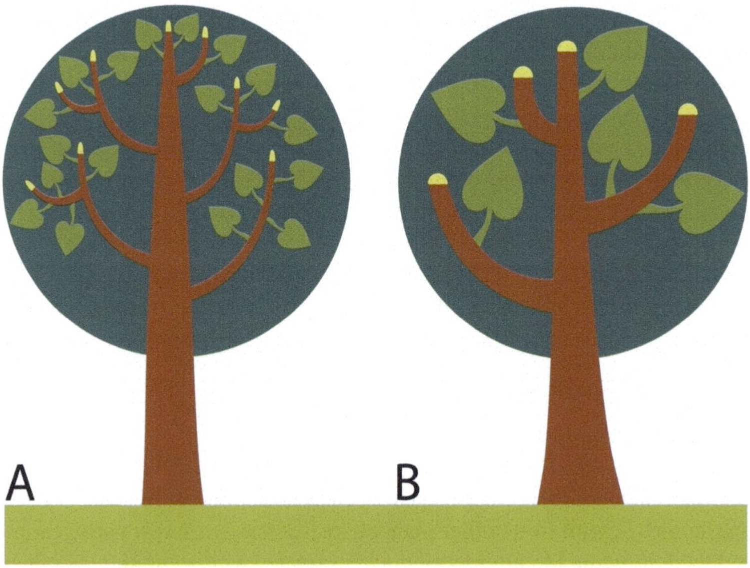

Fig. 3.2 Corner's rule illustrated by two trees differing in the size of the growth apex, leaf size, number of leaves on the shoot, thickness of the shoot, and number of branches in the crown. (**A**) *Tilia cordata*, (**B**) *Paulownia tomentosa*

Even though we have so far mentioned the relationship of shoot properties, described as Corner's rule, only for vegetative organs, it should be noted that it also applies to generative organs: flowers, fruits, and seeds. This is because a large growth apex can transform into a large inflorescence or flower, produce a large number of small fruits or seeds, or really large seeds. A small growth apex is limited in this respect and will always produce a smaller quantity or smaller seeds than a large meristem.

3.1.4 Spatial Arrangement of Leaves

The growth apex produces leaves in a certain spatial and temporal arrangement. Much attention has been devoted to examining the spatial arrangement of leaves (phyllotaxis), especially the description of the spiral arrangement, which exhibits the Fibonacci sequence (Box 14) (a numerical series starting with zero and one, with each subsequent number formed by adding the two preceding ones (Fk = Fk-1 + Fk-2), thus the beginning of the sequence looks like this: 0, 1, 1, 2, 3, 5, 8, 13, 21, 34, 55, 89, 144).

Charles Bonnet (Box 15) was the first to systematically study the arrangement of leaves along the stem. In his book from 1754, he described four different leaf arrangements: opposite leaves, whorls of three or more leaves, alternate leaves on opposite sides of the stem, and leaves without apparent arrangement. By further observing these unordered leaves, he arrived at the following: (Adler et al. 1997) "*Imagine five vertical lines drawn on a cylindrical stem, with equal spaces between them. Put a leaf at the bottom of the first line. Put a second leaf slightly higher up on the third line to the right or the left. Continue in the same direction, putting the third leaf on the fifth line, the fourth leaf on the second line, and the fifth leaf on the fourth line, completing a cycle of five leaves. The sixth leaf, placed on the first line higher than the fifth leaf, starts a new cycle. In this arrangement there are two complete rotations around the stem between the first and sixth leaves, and the leaves between them divide this amount of rotation into five equal parts, so the angle between consecutive leaves is 2/5 of a rotation, or 144°*". Bonnet also provides an explanation for the purpose of the entire phenomenon; it is to minimize the overlap of leaves and thus allow free air circulation.

Scholars connected the Fibonacci sequence with spirally arranged leaves even before Bonnet, and they were not botanists. For example, the astronomer Kepler noticed five leaves around the stem in a spiral (and other pentamerous features in plants) and contrasted it with minerals, where pentamerous features are not found. The Fibonacci sequence was gradually found in many spirally arranged natural phenomena such as the nautilus shell, ocean waves, leaf rosettes, pinecones, or sunflower inflorescences. This trend prompted observers to search further and research phyllotaxis, whether as evidence of divine existence or as an interesting problem in developmental biology, ecology, or evolutionary biology. When the whole issue is greatly simplified, we can say that developmental biologists believe that a new leaf primordium appears periodically at the edge of the growth apex at the smallest possible distance from the previous primordium; evolutionary biologists admit that phyllotaxis could be an adaptation minimizing leaf overlap and thus shading within a single plant, and ecologists emphasize that there are other ways to limit self-shading than phyllotaxis, such as branching, leaf shape plasticity, and petiole length. However, there are rosette plants with rigid leaves that cannot influence leaf sun exposure, and phyllotaxis is important for them. American biologist Karl Niklas and Sean Hammond (2013), using a mathematical model, demonstrated that the greatest sun exposure is provided by the Fibonacci angle between successive leaves of 137.5°, and the effect is more pronounced in narrow leaves because oval leaves overlap more (Fig. 3.3).

3.1.5 Timing and Speed of Leaf Development

Leaf primordia are established at the growth apex always after a certain time, which we call plastochron (the time between the unfolding of subsequent leaves from the bud is called phyllochron). Plastochron is mainly used in developmental biology and in the detailed study of crop physiology and model species. It allows for the

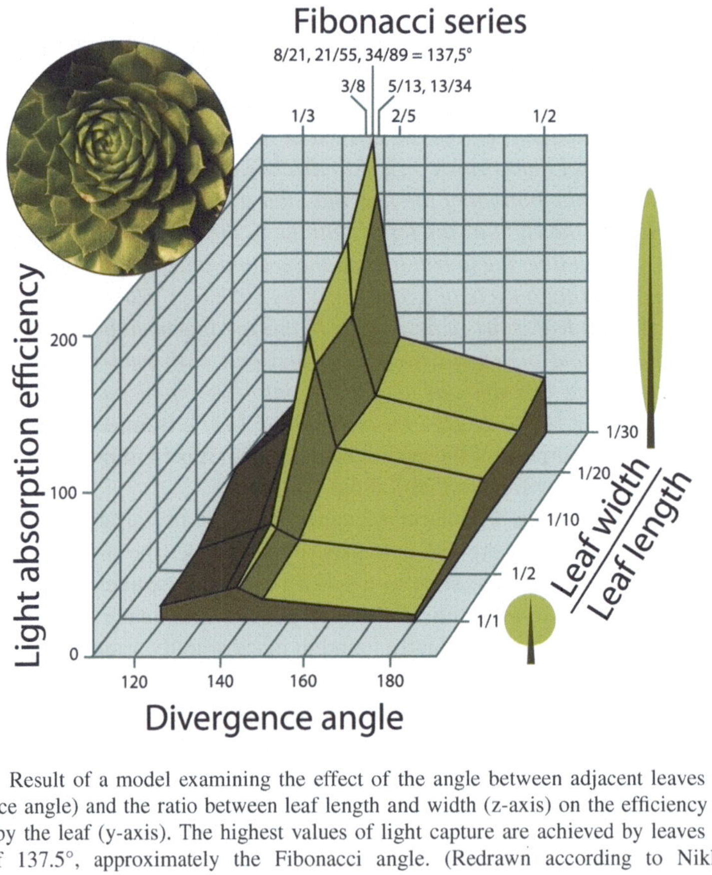

Fig. 3.3 Result of a model examining the effect of the angle between adjacent leaves (x-axis: divergence angle) and the ratio between leaf length and width (z-axis) on the efficiency of light capture by the leaf (y-axis). The highest values of light capture are achieved by leaves with an angle of 137.5°, approximately the Fibonacci angle. (Redrawn according to Niklas and Hammond 2013)

comparison of individuals based on their morphological development, the speed of which depends on external conditions, where comparing individuals of the same age may not make sense. In ecology, for example, plastochron was used to describe the growth of lateral branches of the clonal plant *Glechoma hederacea* (Birch and Hutchings 1992), which depended on the order of nodes on the parent shoot and the order in which the branches grew. This effect can be considered an architectural constraint on plasticity in the species' response to a heterogeneous environment. Certainly, interesting questions could be tested using plastochron in different species, but for many species, we lack data.

Besides the length of individual phases of growth apex development, their timing is interesting. For example, under what conditions does the growth apex transition from leaf production to flower production or from dormancy to growth. The products of the growth apex (leaves and flowers) can emerge from the bud immediately after their establishment (neoformation) or can accumulate in a closed bud and only unfold at an opportune moment (preformation). It can even happen that these products do not emerge from the bud gradually as they were established from oldest to youngest, but in reverse order. An example is *Colchicum autumnalis*, which blooms in the fall and this autumn flower precedes the leaves, which remain hidden in the bud and emerge only in the next season along with the infrutescence (Fig. 3.4).

Preformation and neoformation have been studied in both woody plants and herbs. For example, in the rubber tree, which grows in a tropical climate without marked seasonality, the development of the growth apex is periodic. Inside the bud, leaf primordia first accumulate, then they unfold, giving rise to leaves and branches, and then the bud rests. Periods of activity and rest alternate several times during the season without an apparent external signal, so it is assumed that the growth apex of the rubber tree needs periodicity for its growth (Hallé et al. 1978). In contrast, the growth apices of palms produce leaves continuously and regularly. In seasonal climates, growth periodicity is induced by the external environment, and trees in seasonal climates often have preformed entire annual increments, including inflorescences, which is why they green up and bloom very quickly in spring (Fig. 3.5).

In herbs, preformation of leaves or even entire inflorescences well before the unfolding (before winter) ere are only isolated observations of the bud stored belowground is also known. Thilo Irmisch (Box 16) already noticed it and describes it in his book on tuberous plants from 1850. The Russian botanist Serebriakov (Box 17) also described the timing of flower establishment for a number of species from lime forests around Moscow. Preformation has so far been studied in plants from environments with a very short growing season, such as the understory of deciduous forests, the alpine zone, and the Arctic (for example Diggle 1997). In these environments, preformation before winter allows plants not to waste time in spring but to unfold leaves and flowers as soon as possible to ensure they can fruit. According to an analysis of the flora of the Czech Republic, preformation of shoots with flowers ready to unfold is not rare even here (Schnablová et al. 2020), and a third of herbs are capable of it. In line with published observations, it mainly concerns species that bloom very early in spring (Fig. 3.6), but not only those. For example, *Geranium*

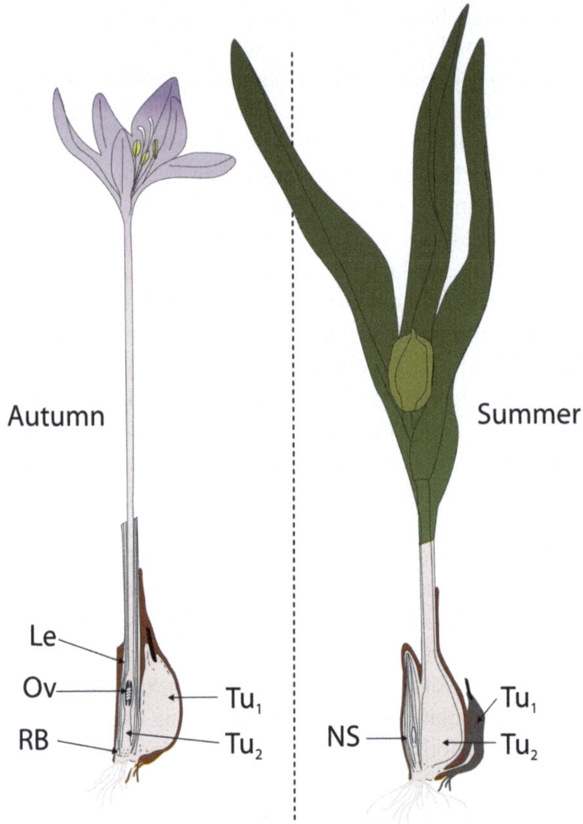

Fig. 3.4 The shoot of *Colchicum autumnale* unfold organs in the reverse order in which they were established at the growth apex. In autumn: the flower, which was established from the apical meristem of the shoot, emerges aboveground, leaves that were established earlier remain undeveloped; in the summer of the following year, the fruit and leaves grow above-ground. Le leaf bases, Ov ovary, RB renewal bud, which continues to grow in summer as a new shoot (NS) and will bloom the following autumn; Tu tuber, Tu_1 parental tuber in autumn, it is a storage organ, in the next summer it is depleted; Tu_2 offspring tuber in autumn, it does not yet have a storage function, in the next summer when the leaves belonging to the same shoot are green, it has the storage function

pratense and *Symphytum officinale* are among the species preforming inflorescences, even though they bloom relatively late.

There are only isolated observations about when exactly temperate zone herbs prepare buds for the next year, according to which it seems that some establish inflorescences in the bud already the previous year in summer or autumn. They can thus avoid low temperatures, at which cell division is significantly slowed down. Due to climate change, plants with preformed buds may have problems successfully blooming and producing fruits. Early snowmelt can cause premature unfolding of preformed flowers, and subsequent late frosts can then destroy them.

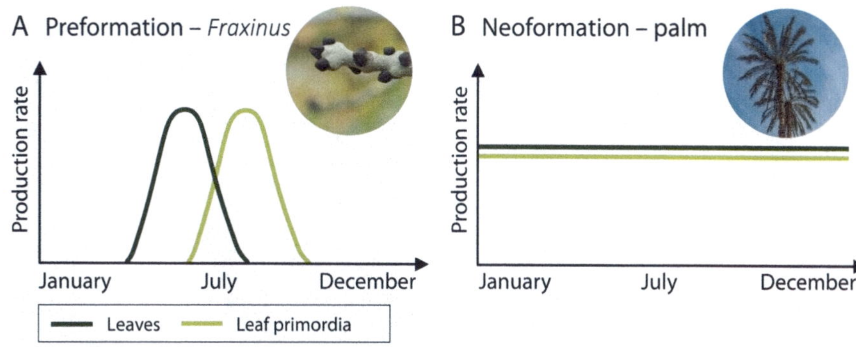

Fig. 3.5 Seasonal development of the growth apex (apical meristem) in trees of the temperate zone (ash, *Fraxinus*) and tropical trees (palm). While the ash accumulates the established beginnings (primordia) of leaves and inflorescences during the previous growing season, the palm directly develops each established organ, and the beginnings of new organs do not accumulate in the bud. (According to Hallé et al. 1978)

3.1.6 Size and Shape of Leaves

According to Corner's rule, leaf size depends on the size of the growth apex. The leaves of the world's flora differ in size by about 100,000 times, and leaf size is often a studied characteristic of a plant (see Sect. 2.2). How do we, as ecologists, explain the significance of leaf size? Since leaves are largest in humid tropical areas and smallest in dry areas, alpine zones, and arctic tundra, an explanation through climate is suggested. Leaf temperature is key for photosynthesis, and thus for the production of organic substances, which increases in efficiency with temperature. At the same time, with increasing temperature, the opposite processes accelerate, during which organic substances are broken down during respiration. Therefore, the optimal efficiency of photosynthesis is in the middle of the temperature gradient, shifted to higher temperatures for plants from warm areas than for plants from cold areas. Plants cannot actively regulate their temperature, but they have ways to ensure that in very warm climates, leaves do not overheat and become ineffective or even destroyed by heat. A shiny reflective surface or orienting the leaf edge instead of the surface towards the sun's rays can be effective strategies where there is not enough water. In well-watered ecosystems, however, a large transpiring leaf is the most effective cooling. Large leaves have a thicker boundary layer, which slows down the exchange of heat between the leaf and the air, meaning that large leaves can maintain a greater temperature difference between the air *versus* the plant than small leaves (Wright et al. 2017). The same mechanism can explain why plants in cold areas have small leaves. The large boundary layer in large leaves at night makes it difficult to exchange heat with the soil and surrounding vegetation, leading to heat radiation and excessive cooling or even the freezing of large leaves.

Temperature regulation on the leaf surface, however, may not be the only reason for a certain leaf size. Large leaf size in the tropics means that plants there have a large apical meristem, which is susceptible to freezing and would not survive in the

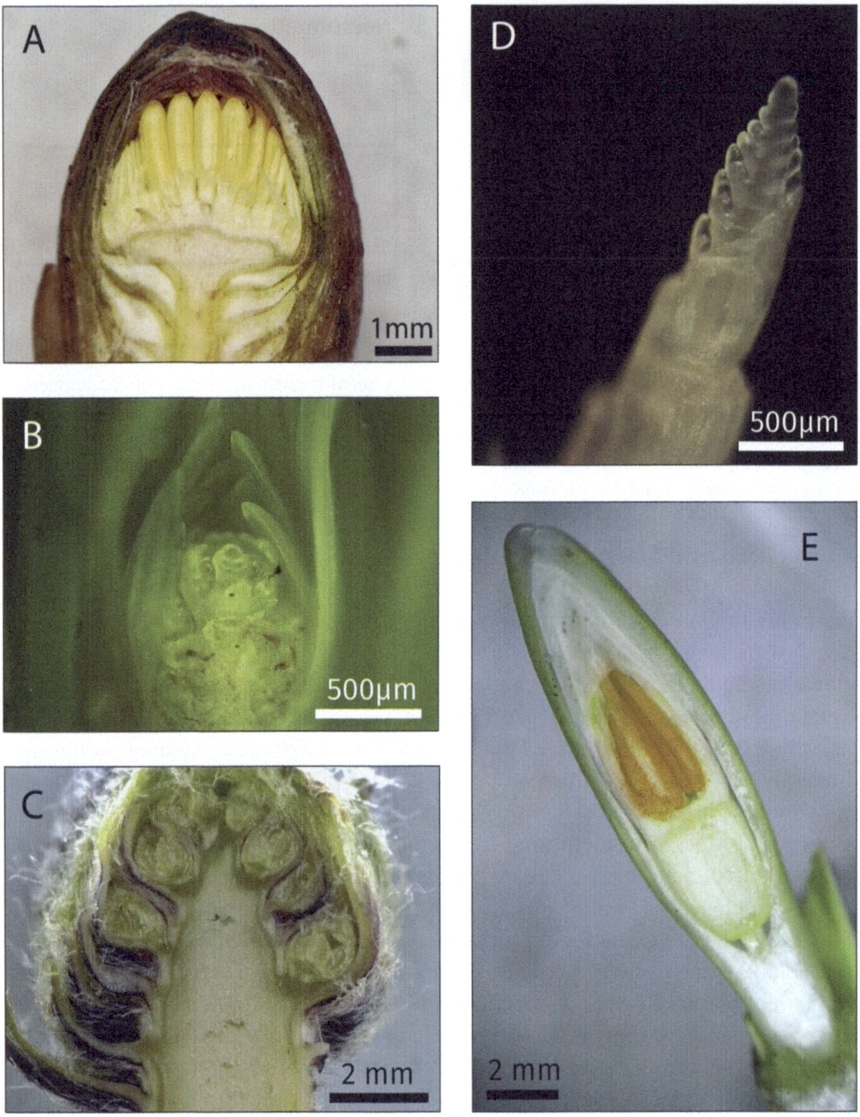

Fig. 3.6 Preformed flowers or inflorescences in belowground renewal buds of herbs before winter. (Photo **A, B, D, E**: Renáta Schnablová) (**A**) *Tusilago farfara*, (**B**) *Euphorbia cyparissias*, (**C**) *Pedicularis dasyantha*, (Photo Alexandr Pospěch), (**D**) *Festuca rupicola*, (**E**) *Galanthus nivalis*

temperate zone. Palms, which have typically large leaves, have only one representative in Europe in southern Spain, *Chamaerops humilis*, which is a small palm with leaves smaller than the temperate herb, *Petasites hybridus*. However, unlike the dwarf palm, *Petasites hybridus* has its growth apex hidden belowground and thus protected from frost. Moreover, large leaves take longer to produce, have a longer plastochron, and in a seasonal climate, there would be little time to create large leaves. The plant would manage only one large leaf per season, which is very risky because what would the plant do if it lost this leaf? Leaf size could also be an evolutionary byproduct of having a bud bank, because the smaller the leaves you produce, the more you can manage in a season, and the more reserve buds you have on the produced stem, from which you can branch and regenerate if needed (see also Sect. 3.4). Very small leaves and richly branched stems can lead to a cushion-like growth form, which is advantageous in windy and cold climates.

Interestingly, there are many plants with compound leaves, which means that the plant has a large meristem, but the individual leaf blades (leaflets of a pinnate leaf) are small. The division of leaves into leaflets could be caused by the fact that the plant cannot afford such a large leaf as would correspond to the size of the meristem, because it does not grow in environments where there is plenty of water. Divided leaves can also serve as protection against mechanical damage from wind or herbivore grazing. Damage to a small area on a large leaf can lead to its loss, but the same damage for a compound leaf may result in the loss of only one leaflet. Many mechanisms and influences on leaf shape have been proposed, but there is no universal theory that explains the shape differences between leaves (Ferris 2019).

3.1.7 Plant Size

As a plant grows, not only does its overall size and the size of its individual organs change, but also its anatomical structure. A plant increases in size during ontogeny in three possible ways: (i) it produces more and more growth modules that remain the same size as in a young plant; (ii) it produces more and more growth modules while primarily thickening, or (iii) it produces more and more growth modules while thickening secondarily. In the latter two cases, the size and anatomical structure of the organs of young and old plants differ. The subsequent description is a certain simplification of rather complex phenomena; for those interested in details, I refer to specialized literature, e.g. (Lux et al. 2017).

The first example, where nothing changes in the anatomical structure, is relatively rare and mainly concerns small short-lived plants that flower very early. An example can be the smallest terrestrial plant, an annual found in the Arctic, *Koenigia islandica*, or small creeping and adventitiously rooting plants from the genus *Callitriche*, growing on the exposed mud of seasonal ponds.

Primary thickening occurs by enlarging the growth apex (see Corner's rule) to a size typical for the species. You have surely noticed that seedlings have thinner and lower shoots with smaller leaves than the adult plant. By increasing the number of undifferentiated cells in the growth apex, the plant is capable of producing a thicker stem with larger leaves. While dicotyledonous plants combine primary thickening when young with the enlargement of the perennial parts of the stem and root through secondary thickening, monocotyledonous plants rely only on primary thickening (though there are exceptions here too). Primary thickening affects the thickness of newly produced shoots, but secondary thickening thickens stems or roots that have already grown (Fig. 3.7). Primary thickening mainly concerns stems, but secondary thickening also concerns roots. Thus, a dicotyledonous plant has thick stems and roots, while a monocotyledonous tree, like a palm, has only a thick stem, but the roots are relatively thin.

Primary thickening enables herbaceous plants that have to reach their maximum possible vertical size each year anew (in seasonal climate) to produce taller shoots. But increasing meristem size through primary thickening has its limits in seasonal climates, as we discussed elsewhere (see Sect. 2.1.3). The additional not yet discussed reason may be the heaviness of such large leaves, as might be inferred from the fact that the stems from which they are growing are not elevated above the soil surface (for example in *Petasites hybridus*). Secondary thickening, on the other hand, enables woody plants to attain their maximum vertical size through lignified bodies that persist over years. Although treeish monocotyledonous plants like palms may be tall without secondary thickening, their stem has different mechanical properties than the wood of angiosperms (Niklas 1992).

During primary thickening, leaves change due to the enlargement of the growth apex. In plants with significant primary thickening, the size difference of the leaves between the seedling and the adult plant is greater than in plants with secondary thickening. Leaves can also change their shape, size, phylotaxis, longevity, photosynthetic efficiency, disease and insect resistance, or epidermal features (heteroblasty) (Jones 1999). Leaf shape in young plants is usually simpler, the leaves are

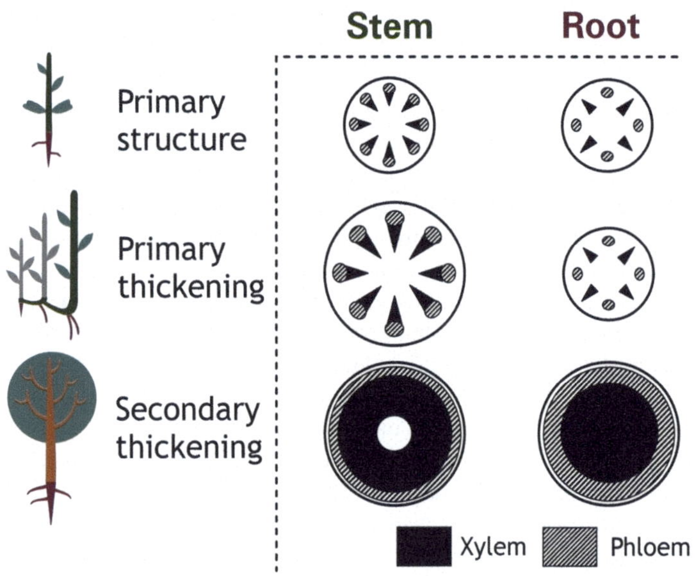

Fig. 3.7 Changes in the anatomical structure of the stem and root of dicotyledonous plants depending on whether they thicken primarily or secondarily (compared to the primary structure in the first row). Xylem—conductive tissue carrying water and minerals from the roots to the leaves, Phloem—conductive tissue carrying organic substances from the leaves to the roots; storage tissue, parenchyma, is marked in white

less divided and have fewer leaflets in the case of compound leaves. Leaves survive the entire life of a perennial plant only exceptionally; this occurs in short-lived plants, such as annuals, or plants with special architecture that cannot establish new leaves during their lifespan (see the chapter on Architectural Peculiarities). The anatomical structure of the leaf usually does not change during leaf or plant lifespan but depends on leaf position in the plant canopy.

3.2 Architectural Models

3.2.1 How It Began

So far, we have only discussed some parts of architecture, such as modules and meristems (buds, growth apices), but now we will look at how the addition of more and more generations of shoots occurs during the life of a plant. How did the view of the plant through its repeating modules (and not through the shapes of the stem, root, and leaf) originate? Both approaches are derived from Goethe's work on the metamorphosis of plants. While comparative morphology analyzes how individual organs adapt their shape to their function (see analogous and homologous organs), architecture reconstructs how subsequent generations of shoots grow (in trees) or are replaced (in herbs). From this difference, it follows that comparative morphology is more concerned with the function of plant parts but is inherently a rather static discipline; architecture is concerned with the dynamics of growth but rarely seeks to connect recognized growth types with their ecological function. And that is, of course, a pity, because there lies great potential, and I hope that the following text will be sufficiently inspiring for ecologists.

The first to systematically focus on plant architecture, in this case, herbs of the temperate zone, was Thilo Irmisch (Box 16), who dealt with the ontogenetic and seasonal development of plants. His work was based on a detailed description of individual species (Irmisch 1850) and did not reach any generalization or classification (or at least he did not publish it). Another prominent figure in plant architecture is Werner Rauh (Box 18). Rauh focused on groups of plants that are not particularly interesting from the perspective of comparative morphology (they have completely normal leaves, stems, and roots) but are interesting precisely because of their architecture, such as cushion plants (Rauh 1939), espalier shrubs (Rauh 1937a), or root-sprouting species (Rauh 1937b). He had plants sent to him, cultivated them, and observed their growth. He did not stop at descriptions of individual species; on the contrary, he always created a classification of growth for the studied group, the first architectural models—a summary of branching rules, leaf arrangement along the shoot, their persistence, etc., which he illustrated in schematic drawings (Fig. 3.8). The classification of growth models in his work was always tailored to the studied growth form; he did not attempt to generalize beyond individual growth forms. Later he focused more on taxonomy and hung-up plant architecture. However, his pre-war research, especially on the classification of growth in temperate woody

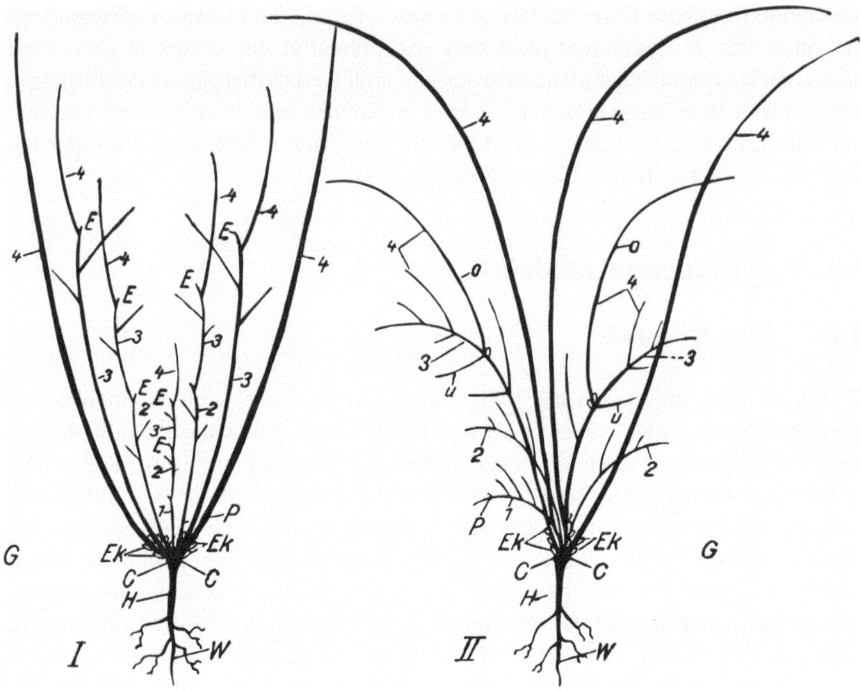

Fig. 3.8 Rauh's classification of branching in European shrubs. I—acrotonic branching (from the upper buds of the shoot), examples: *Corylus avelana*, *Evonymus europea*, *Rhododendron*; II—mesotonic branching (from the middle buds of the shoot), examples: *Berberis vulgaris*, *Philadelphus coronarius*, *Forsythia suspensa*. Arabic numerals indicate the generations of shoots as they gradually grow, H hypocotyl, W main root, C cotyledons, Ek epicotyl (first segment of the stem). (Redrawn from Rauh 1937a)

plants, was followed by morphologists worldwide, particularly those studying the growth of woody plants in the tropics, where the center of diversity of tree forms lies.

3.2.2 Trees

Accumulating knowledge about the growth and architecture of tropical trees culminated in the 1970s in a synthesis that made plant architecture a new field. French botanist Francis Hallé (Box 19), who studied tree growth and also lived for several years in tropical countries in Africa, Indonesia, and South America, published a book on the architecture of tropical trees in 1970 together with Dutch botanist Oldeman, in French. After including American researcher Tomlinson in the authorship team (henceforth HOT), a new and expanded edition of the book was published in English in 1978 (Hallé et al. 1978). The book has been frequently cited since its publication and is required reading for botanists and foresters working in tropical forests. The diversity of tree growth is brilliantly synthesized into 23 architectural models named after significant morphologists (Corner, Rauh, and Troll each have their model here). HOT builds on Rauh's approach, but by working in the tropics, where lays the center of tree diversity, their classification is truly global.

The authors consider any plant that looks like a tree, meaning that it is large and has a trunk and crown, to be a tree. Trees we are accustomed to from the temperate zone have a trunk and crown from the beginning of life, both of which just increase in size, while palms or cycads, for example, first have a crown (in this case, a tuft of leaves) on a very short trunk, which gains only length over time. The classification is based on whether the plant has only a single axis or if it has more, whether axes are equivalent or different, whether they branch sympodially or monopodially and regularly or irregularly, and so on (Fig. 3.9). For classifying a tree in the HOT system, we need to describe the tree's growth under optimal conditions (for example, in a botanical garden or plantations) until its first flowering. In the real forest environment, the architectural model tends to manifest, but it is influenced by growth conditions such as competition, disturbance, or shading. Due to environmental influences and disturbances, there is the possibility of a reduction or reiteration of the basic models.

3.2.3 Herbs

The architectural models of HOT aim to include all plants without distinction, and the book contains a chapter on herbs, claiming that herbs grow according to the same models, only they have undergone miniaturization, fragmentation, or reduction of the model. A manuscript thesis from the University of Montpellier is cited, in which the issue of architectural models in herbs is addressed. However, no one has ever followed up on these notes, and the application of HOT models to herbs has never been developed. Nevertheless, there exists, thanks to a publication in Russian, a completely hidden work that was inspired by HOT models but is tailored to herbaceous species. The author of this work is T. I. Serebriakova (Box 20), who in 1976 (thus before the publication of HOT in English) published a short communication in the journal "Moscow Researchers of Nature" about the architectural models of herbs (Serebriakova 1977). Serebriakova either encountered the first edition of HOT or heard a lecture by Francis Hallé at the 1975 botanical congress in Leningrad (today's St. Petersburg), which inspired her classification.

Serebriakova's categories for all herbs are only 4 and are based on the fate of the primary shoot and its branching; they are not named after morphologists but are numbered (Fig. 3.10). Her division was rarely cited in Russian literature and was not used in plant ecology (which also applies to architectural models according to HOT). The disadvantage is, as with the HOT classification, that it overlooks root sprouting, but focuses only on the architecture of the plant based on stem branching and leaf arrangement along the stem.

The main feature for distinguishing categories is the type of stem branching: sympodial or monopodial. In plants with sympodial branching, the growth apex after a shorter or longer period of leaf production transforms into a flower or

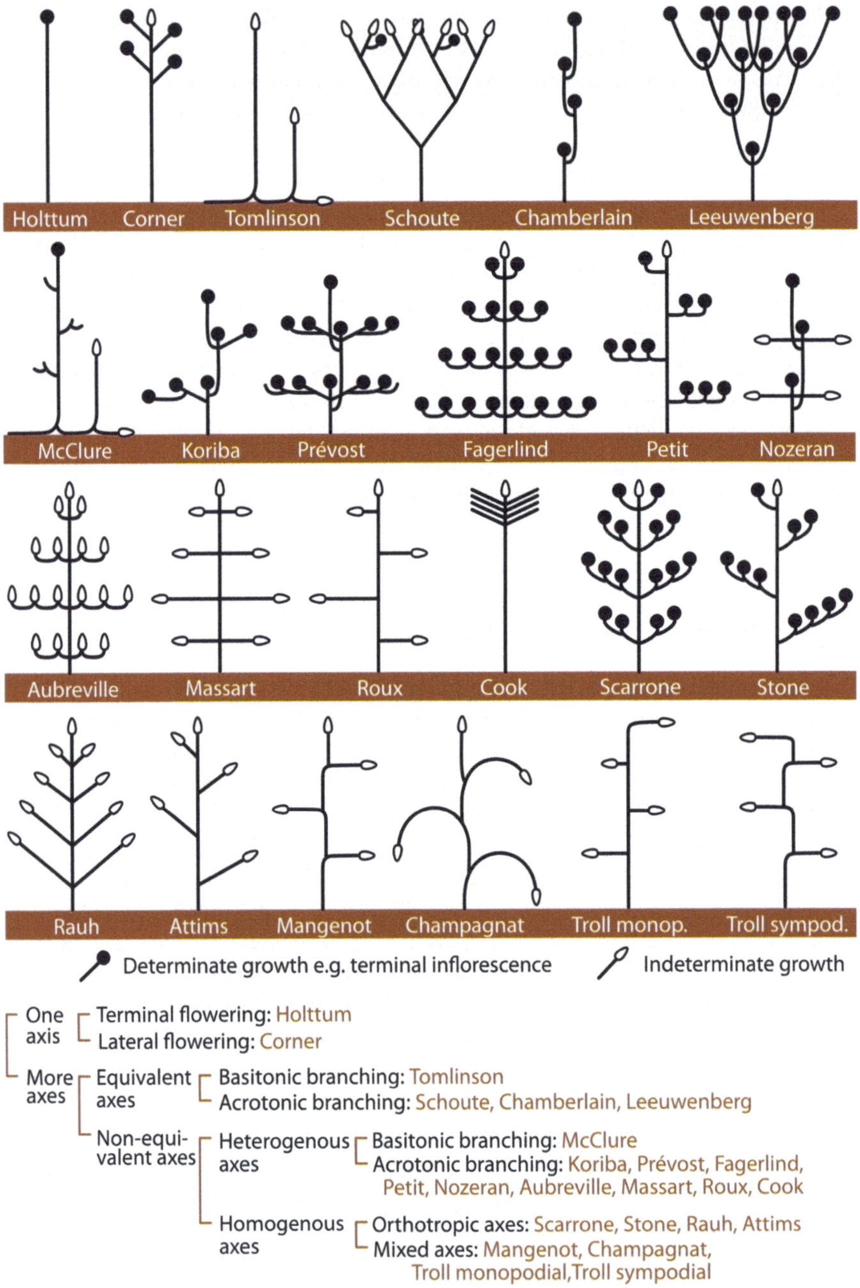

Fig. 3.9 Classification of tree architecture according to HOT and a simple key for determining the main groups. Axis—trunk and branches regardless of how many shoots they originated from; basitonic branching—branches grow from the lower buds of the shoot, acrotonic branching—branches grow from the upper buds of the shoot. (According to Hallé et al. 1978)

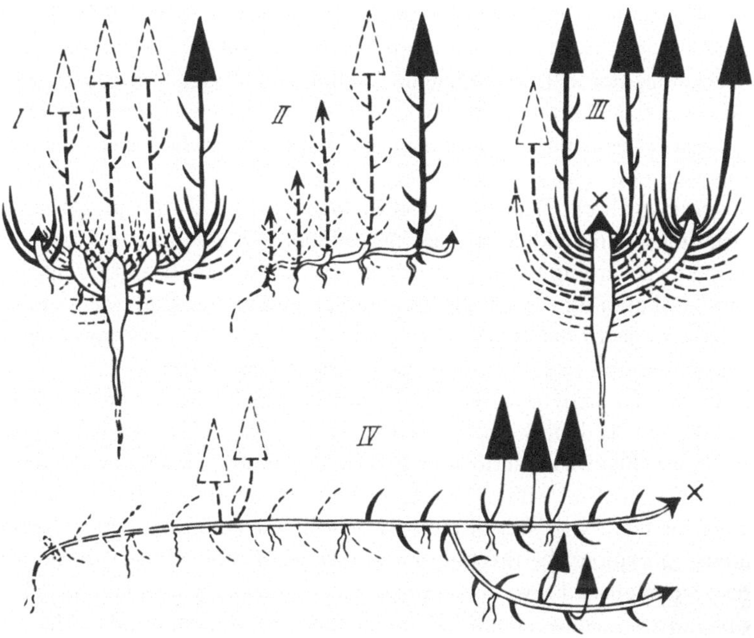

Fig. 3.10 Classification of herb architecture according to Serebriakova (Serebriakova 1977). I—sympodial branching, polycyclic semi-rosette shoots, II—sympodial branching, monocyclic non-rosette shoots, III—monopodial branching, rosette/semi-rosette shoots, IV—monopodial branching, non-rosette shoots. (Redrawn from Serebriakova 1977)

inflorescence, and the plant has one type of shoot: generative (Fig. 3.10, I, II). Plants with monopodial branching have the growth apex in a vegetative state, therefore it produces only leaves and does not differentiate into flower organs (cross on Fig. 3.10, III, IV). The plant blooms from axillary buds so that it produces single flowers, entire inflorescences, or flowering shoots—thus it has two types of shoots: vegetative and generative (Fig. 3.10, III, IV). Both sympodially and monopodially branching plants can form shortened internodes and a leaf rosette at the shoot base (Fig. 3.10, I, III) or only long internodes and shoots without a leaf rosette (Fig. 3.10, II, IV).

3.2.4 Root Systems

Inevitably, we have returned to the roots, which researchers studying plant architecture tend to overlook. However, roots, besides fulfilling their duties belowground in acquiring water and mineral substances and anchoring the plant in the ground, have their own architecture and also influence the architecture of the above-ground parts, as well as being affected by architecture of the above-ground parts.

The root systems of herbs were studied by Lore Kutschera (Box 21), an Austrian botanist who began their examination in weeds. On soil exposures, she described root systems together with her colleague Erwin Lichtenegger, who illustrated them. A unique collection of portraits of root systems of weed flora was created (Kutschera 1960), to which books on the root systems of grasses and other monocotyledonous herbs followed (Kutschera and Lichtenegger 1982), as well as on the root systems of dicotyledonous herbs (Kutschera and Lichtenegger 1992). Besides descriptions and illustrations of individual species, the authors also classified the root systems found in herbs according to the presence of adventitious roots and storage organs, and their distribution in the soil profile. Similar, although not as extensive, studies of the root systems of individual herb species also exist for some other regions or types of communities (e.g., for American prairies (Weaver 1958)).

If the research of herb root systems is demanding and therefore rarely conducted, then for trees, this is doubly true. Nevertheless, there is a classification of root system architecture that was described with the contribution of the Czech ecologist Jan Jeník (Box 22). He dealt with root systems in his dissertation at the forestry faculty, and during his three-year stay at the university in Ghana in the 1960s, he had the opportunity to explore the diversity of woody plant root systems in the tropical rainforest. He published several contributions on interesting adaptations of tropical woody plant root systems (Jeník 1973) that supplemented and refined the existing Krasilnikov classification (Krasilnikov 1970). He thus contributed to the knowledge of tropical tree roots at a time when HOT architectural models were relatively fresh from the press, and his contribution was published in a collection edited by Tomlinson in 1978 (Jeník 1978) (Fig. 3.11).

Jeník distinguished tree root systems into 23 categories according to the following characteristics: presence of secondary thickening, overall structure of the root system of adult individuals, relationship of soil and aerial root subsystems, changes in roots during life, occurrence of abnormal shapes, and modifications due to stress conditions. The types of root systems are named after their typical representatives (Fig. 3.11). The classification remains unused in comparative ecological works, although individual types and their functions are further examined from a functional perspective.

3.2.5 What Do We Know About Evolution?

The architecture of plants is very poorly reconstructed from the fossil record because plant bodies are found in parts, and it happens that roots, stems, and leaves belonging to one individual are often described as separate species. Nevertheless, there are situations in which the reconstruction of the entire plant is possible, and based on these cases, architectural models for spore plants have recently been described according to the principles of the HOT system (Fig. 3.12). The result is 12 growth models (named in honor of paleontologists and morphologists) (Chomicki et al. 2017), which have no counterpart in seed plants because they are based on dichotomous branching, whereas the branching of seed plants is lateral from axillary buds.

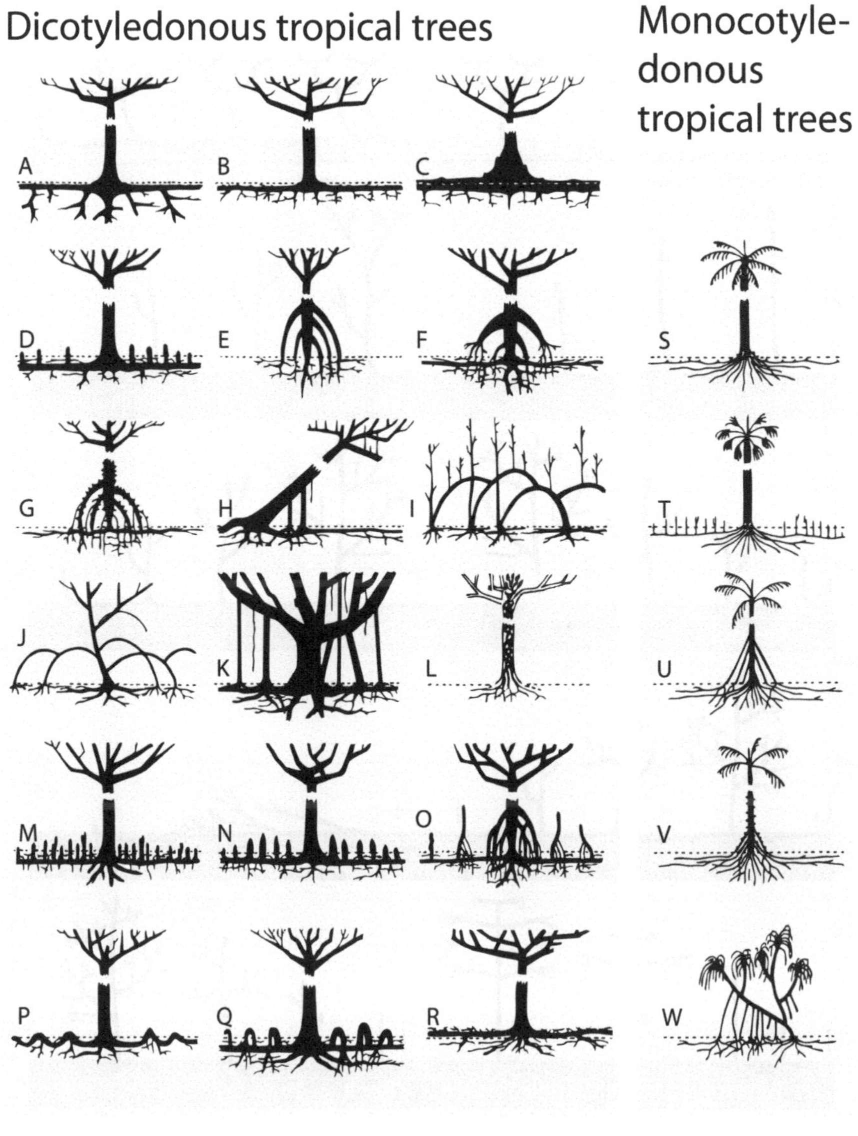

Fig. 3.11 Main organizational models of root systems of tropical trees from the work of Jan Jeník (Krasilnikov 1970), named after typical representatives. (**A**) *Chlorophora excelsa*, (**B**) *Cariniana pyriformis*, (**C**) *Piptadeniastrum africanum*, (**D**) *Xylocarpus mekongensis*, (**E**) *Uapaca guineensis*, (**F**) *Tarrietia utilis*, (**G**) *Bridelia micrantha*, (**H**) *Protomegabaria stapfiana*, (**I**) *Scaphopetalum amoenum*, (**J**) *Rhizophora mangle*, (**K**) *Ficus benjamina*, (**I**) *Ficus leprieuri*, (**M**) *Avicennia germinans*, (**N**) *Sonneratia alba*, (**O**) *Xylopia staudtii*, (**P**) *Bruguiera gymnorrhiza*, (**Q**) *Mitragyna stipulosa*, (**R**) *Alstonia boonei*, (**S**) *Elaeis guineensis*, (**T**) *Mauritia flexuosa*, (**U**) *Socratea exorrhiza*, (**V**) *Crysophilla aculeata*, (**W**) *Pandanus candelabrum*. (Redrawn from Jeník 1978)

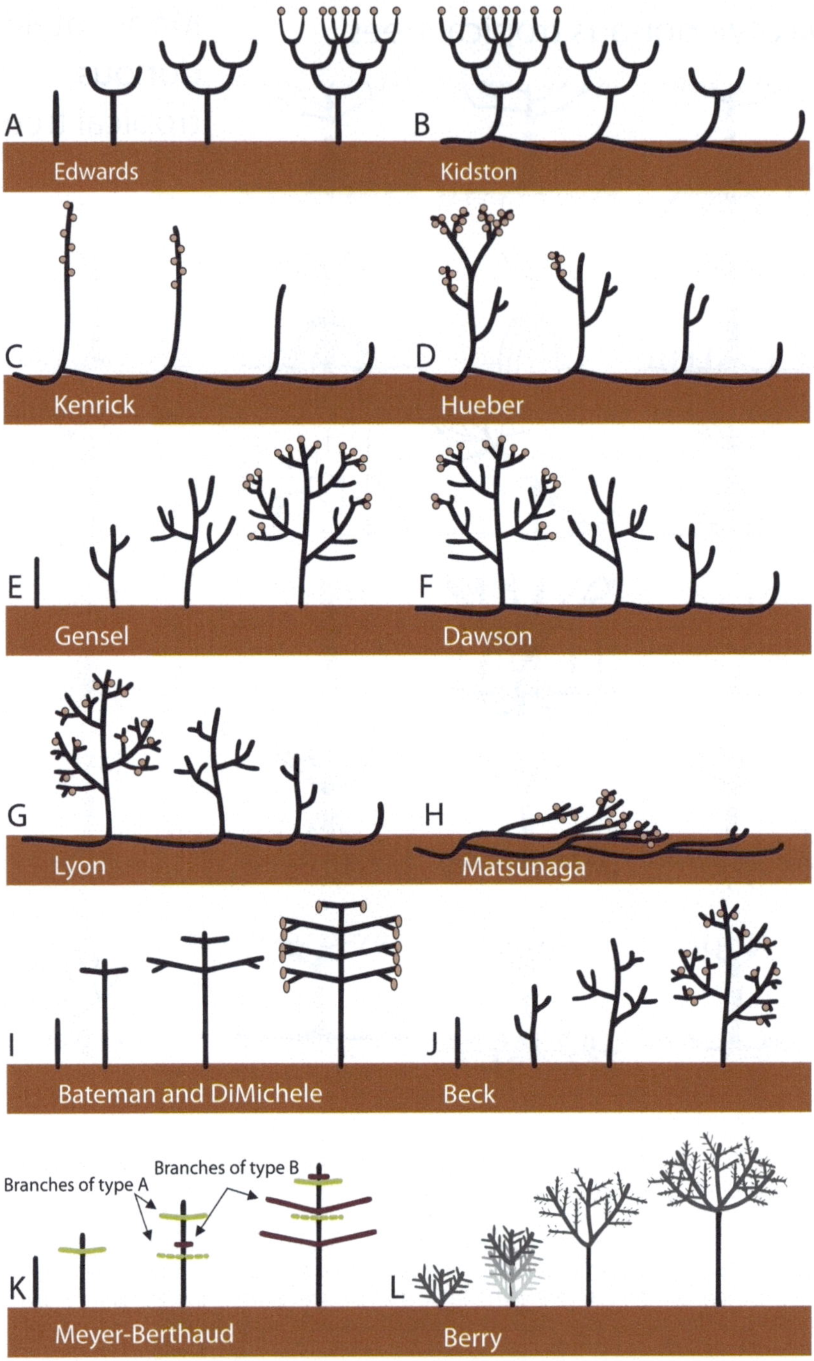

Fig. 3.12 Architectural models of extinct spore plants named after paleontologists and morphologists. (Redrawn from Chomicki et al. 2017)

Spore plants had active meristems only at the branch tips and did not have axillary buds. Branching occurred such that the growth tip enlarged and then divided into two. Stems could thus branch only during their growth, but not later, and therefore did not have a bud bank.

The authors (Chomicki et al. 2017) also reconstructed the transition from dichotomous to axillary branching in the fossil record. This occurred during the Mesozoic era when spore plants were replaced by seed plants. This change was not merely a replacement of one set of architectural models with another in terms of the plant's form and function, but it signified the emergence of a bud bank, thereby providing greater growth flexibility and also greater flexibility in the plant's response to disturbance. Dichotomous branching can also be found on rare occasions in seed plants, mostly in tree-like monocotyledons and cacti, that is, in forms where primary thickening prevails and the lateral meristem (responsible for secondary thickening) is absent, and axillary buds are also missing. Thus, dichotomous branching in these groups is their only method of (rare) branching.

3.3 Relationships Between Shoots and Roots

3.3.1 Primary and Adventitious Roots and Shoots

It is interesting that plant experts, the creators of the architectural model classifications presented in the previous chapter, did not mind that their models only deal with half of the plant, as roots and shoots are not separate entities but are interconnected into one plant. When we return to the beginning of the growth of a seed plant, to the seedling stage, we see that crucial events for the architecture of the entire plant occur at the very start of the plant's life. From the embryo of the germinating seed, two growth apices emerge, the root and the stem, each growing in opposite directions. Subsequently, they form the primary shoot and primary root (see also above). This growth method is called bipolar and is mainly encountered in dicotyledonous plants. Monocotyledonous plants usually have a primary root for a very short time, almost immediately after germination it is supplemented by adventitious roots growing from the stems. Here, the stem determines the plant's growth, and this growth is secondarily monopolar. Why secondarily? In evolution, monopolar growth is older and more original, and it is found in all recent representatives of spore-bearing plants; in extinct representatives, bipolar growth also occurred and was typical for the large woody growth forms of horsetails and clubmosses (Mogie and Hutchings 1990).

During ontogeny, not only the primary root but also the primary shoot can be spontaneously lost and replaced by adventitious shoots from the roots. Although we have so far mainly been interested in the undisturbed growth of the plant when defining architectural models, when you examine plants in nature, you will find that many of them have been injured during their lifetime, and the loss of the primary shoot or primary root has affected their architecture. This fact led morphologists to divide plants according to the role that adventitious rooting or adventitious sprouting plays in the plant's life, forming groups that only produce adventitious roots or shoots after disturbance (regenerative); those that do so spontaneously but their life does not depend on it (additive), and finally those that are dependent on this ability because their primary axes are spontaneously lost and otherwise they would not be able to complete their life cycle (essential).

The categories "essential, additive, and regenerative" were first used by Werner Rauh, who studied in detail sprouting from plant roots. Aeschimann and Bocquet (1980) then applied them also to rooting. Additionally, they combined this concept with the concept of alorhizy and homorhizy by the German morphologist and physiologist Karl Goebel (1855–1932), who distinguishes whether a plant has one dominant root (primary root—alorhizy), or all roots are equivalent (adventitious roots replace the primary root—homorhizy). Analogous to the root system, they created similar categories for the shoot system: when one primary shoot dominates in plants, they speak of alocauly, and when all shoots are equivalent, meaning that adventitious shoots replaced the primary shoot, they speak of homocauly (Fig. 3.13). Unfortunately, in their conception, the types of shoot and root systems stand side by side, and the authors do not address in what different combinations may occur.

The classification of the production of adventitious organs into three categories according to the conditions under which they occur (essential, additive, and regenerative) seems to be very useful, but it requires repeated observation of a single or different species under various conditions, or experimental verification. The production of adventitious shoots and roots is a variable matter and, besides disturbance, can be triggered by other environmental conditions. In the case of adventitious rooting, a triggering factor can be the moisture of the substrate. The same species may produce a number of adventitious roots and grow clonally in a moist habitat, while in a dry habitat, it relies on the main root and does not grow clonally. In root-sprouting species, there is greater production of adventitious shoots in plants growing on a new substrate, such as alluvium or a deposit, where there is little competition from surrounding plants, than in plants growing in a dense meadow stand. These three categories of the significance of adventitious shoots and roots for the life of the plant provide us with guidance on what to observe if we want to define, for example, a clonal plant, or supplement and synthesize the classification of plant architecture for herbs and woody plants.

The occurrence of growth polarity and the ability to produce adventitious shoots or adventitious roots is related to the ease with which plants can grow clonally. A plant that invests in the growth and persistence of the primary stem and root (it is usually a plant capable of secondary thickening) will likely remain non-clonal throughout its life, while a plant that invests only in stem growth and forms new adventitious roots as needed has a greater chance of becoming clonal, i.e., breaking up during its life into several physically separate rooting units. Groff and Kaplan (1988) precisely defined when this can happen: it is when the plant is capable of producing either adventitious roots (i.e., roots growing from the stem) or adventitious shoots (shoots growing from the root) or both (Fig. 3.14). Only in this way can repeated connections (rooting units) arise between the parts of the plant acquiring above-ground resources (shoots) and the part acquiring belowground resources

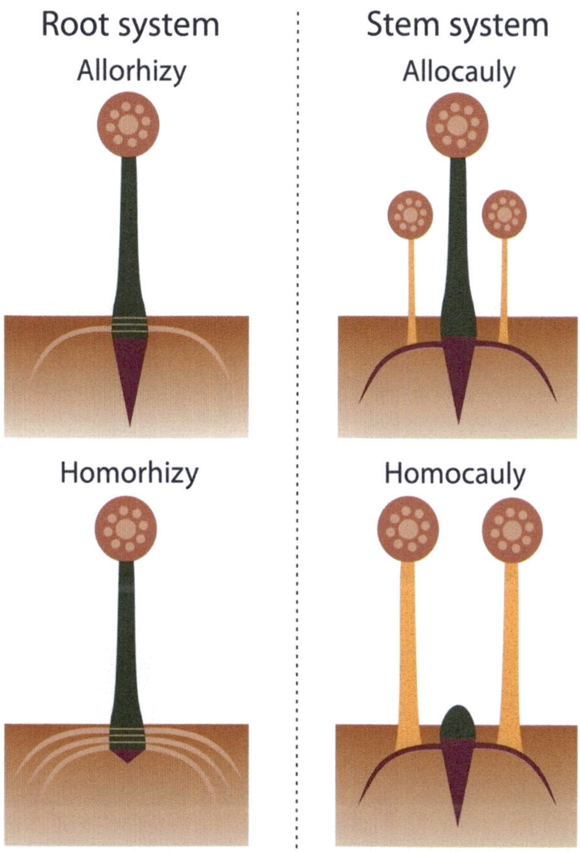

Fig. 3.13 Relationships between roots and shoots according to Aeschimann and Bocquet (1980). The system is based on whether the roots/shoots are equivalent (homorhizy and homocauly), or one is dominant, and the others are subordinate (allorhizy and allocauly)

(roots). As always, there are exceptions that do not fit into this scheme, such as rootless aquatic plants that absorb nutrients from the water through the surface of their bodies and thus are not dependent on roots. In this classification, unlike the classification of Aeschimann and Bocquet, it is not considered whether the formation of adventitious organs is spontaneous or conditioned by injury but it takes into account the combinations between above-ground and belowground organs that can occur in nature.

Plants are not evenly divided into these categories, and the distribution likely differs between biomes and world floras. In Central Europe, about half of the plants are not capable of producing either adventitious roots on stems or adventitious buds on roots, the remaining 40% can create adventitious roots, and only 10% adventitious stems. Of those 10%, an unknown part also forms adventitious roots. We do not have such detailed information about other parts of the world, but we can at least look at the ratio of clonal to non-clonal plants. According to the literature, it seems that in China (Ye et al. 2014), the ratios are similar—about half of the plants are clonal, on the other hand, in Australia, only about 17% of plants are clonal (Zhang et al. 2018). The problem with these data is that we are not entirely sure of the

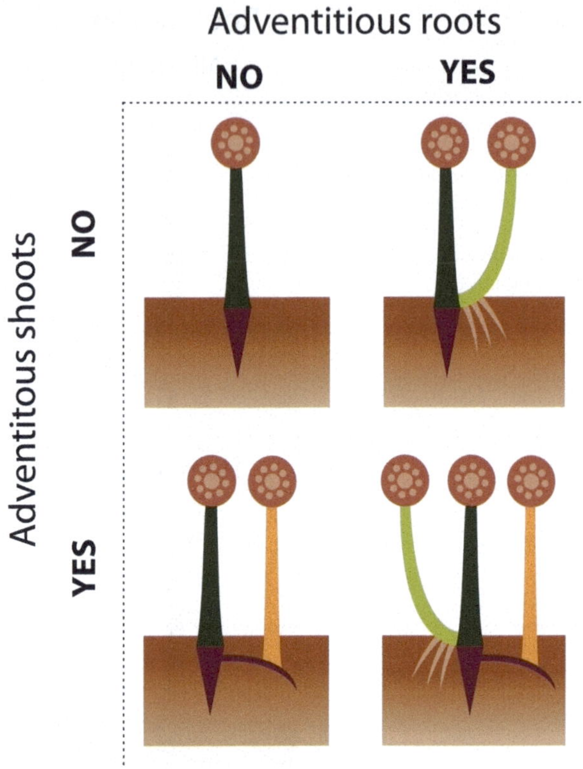

Fig. 3.14 Possible combinations of the arrangement of the relationship between the root and shoot systems of the plant according to Grof and Kaplan (1988). The plant may have only a primary shoot and root, which can further branch and produce axillary shoots and lateral roots (top left). However, the plant can also form adventitious roots (roots growing from the stem; see the second column), adventitious shoots (shoots growing from roots; see the second row), or both, adventitious roots and adventitious shoots (bottom right)

criteria researchers used to define the categories of clonal and non-clonal plants, as they forgot to include it in the description of methods.

3.3.2 Synthesis of Architectural Models

The approaches of HOT and Serebriakova to architectural models are compatible and can be merged, albeit at the cost of significant simplification of the categories according to HOT. I also attempted this (Klimešová 2018) and simultaneously supplemented the models that were missing, namely root sprouting. The resulting 5 categories encompass the main diversity of plant architectural models (Table 3.1, Fig. 3.15). The first category is for plants that manage with the primary shoot for their entire life. The second category is for those in which the primary shoot grows throughout life, but axillary shoots also contribute to the body structure and flowering. The third category includes plants where the primary shoot is part of the plant body only in young plants but is gradually completely replaced by axillary shoots. The fourth category consists of plants where the primary shoot is replaced by adventitious shoots from the roots, and the fifth category includes plants with a reduced body structure that do not have stems (see architectural peculiarities). The first four

Table 3.1 Comparison of synthetic classification and classifications of architectural models for herbs and trees

Type (Fig. 3.15)	Description	Herbs	Trees
1	The plant has only a primary shoot for its entire life, the growth apex does not flower	Creeping perennial herbs, partially category IV according to (Serebriakova 1977), some annuals according to (Krumbiegel 1998)	Monopodial palms, Corner model according to (Hallé et al. 1978)
	The plant has only a primary shoot for its entire life, the growth apex flowers	Most annuals and biennials according to (Krumbiegel 1998; Krumbiegel 1999)	Monocarpic palms, Holttum model according to (Hallé et al. 1978)
2	The primary shoot grows vegetatively throughout the plant's life, axillary branches flower	Upright perennial monopodial herbs, category III according to (Serebriakova 1977); creeping monopodial perennial herbs, partially category IV according to (Serebriakova 1977); monopodially branched annuals according to (Krumbiegel 1998)	Trees with monopodially branched trunks, models Fagerlind, Petit, Aubréville, Massart, Roux, Cook, Scarrone, Stone, Rauh, Attims, and Troll1 according to (Hallé et al. 1978)
3	The primary shoot is part of the adult plant's body but does not grow throughout its life, axillary branches flower	Non-clonal perennial herbs, category I according to (Serebriakova 1977), categories 11 and 14 according to (Klimešová and Klimeš 2008)	Trees with sympodially branched trunks, models Schoute, Chamberlain, Leeuwenberg, Koriba, Prévost, Nozeran, Mangenot, Champagnat, and Troll2 according to (Hallé et al. 1978)
	The adult plant body consists only of axillary shoots, the primary shoot and root are replaced	Clonal perennial herbs, category II according to (Serebriakova 1977), categories 2–10, 12, 13, 16, and 17 according to (Klimešová and Klimeš 2008)	Trees and tree-like herbs, models Tomlinson and McClure according to (Hallé et al. 1978), category I according to (Jeník 1978)
4	The adult plant body consists of adventitious shoots growing from the root (which may further have axillary shoots)	Perennial herbs, category 15 according to (Klimešová and Klimeš 2008)	Root-suckering woody plants, category E according to (Jeník 1994)

(continued)

Table 3.1 (continued)

Type (Fig. 3.15)	Description	Herbs	Trees
5	The primary shoot does not develop at all, adventitious shoots from roots or hypocotyl (usually without axillary shoots)	Mycoheterotrophs and parasites, according to (Rauh 1937b), carnivorous plants from the family Lentibulariaceae	Not found in trees
	The plant does not form primary shoots, lacks stems, and inflorescences grow adventitiously from the hypocotyl, cotyledon leaves, or roots	Podostemaceae, Streptocarpus	
	Other peculiarities	Lemnaceae	

According to (Klimešová (2018)

categories can be further divided based on whether the axillary shoots branch along the stem or from its base. In cases where branching is from the base, we distinguish if the bases of the stems are capable of forming adventitious roots, making the plants clonal (Fig. 3.15—II and III).

A more detailed classification of temperate herbs can be obtained by dividing them into groups according to the distribution of leaves along the stems and branching. Herbs whose shoots form long segments and have a leafy stem without a basal rosette of leaves simultaneously have these as annual shoots (monocyclic, because they live only one growing season from bud development into a shoot until it flowers) because the above-ground parts of the shoots are unable to survive the winter. Species that produce shoots with shortened segments and a basal rosette of leaves simultaneously have perennial shoots (di- and polycyclic), because they can form only a leaf rosette with a growth apex at the soil surface in the first year(s) of life, thus avoiding freezing and then flowering the following year. Monocyclic non-rosette shoots can potentially flower every year but they are more demanding in terms of nutrients and more sensitive to disturbance because when regenerating from a bud bank, they again produce monocyclic shoots, which mostly do not manage to fruit by the end of the season or miss the signals needed for flower initiation. Polycyclic semi-rosette and rosette shoots can accumulate energy for flowering over several years and when disturbed only lose leaves and not the growth apex. They also do not require as nutrient-rich an environment for growth as plants with monocyclic shoots (Klimešová and Klimeš 2008). When in each of the architectural categories (except category 5, which includes plants without shoots) we distinguish the cyclicity of the shoot, we obtain a classification that covers the growth diversity of temperate herbs (Fig. 3.16).

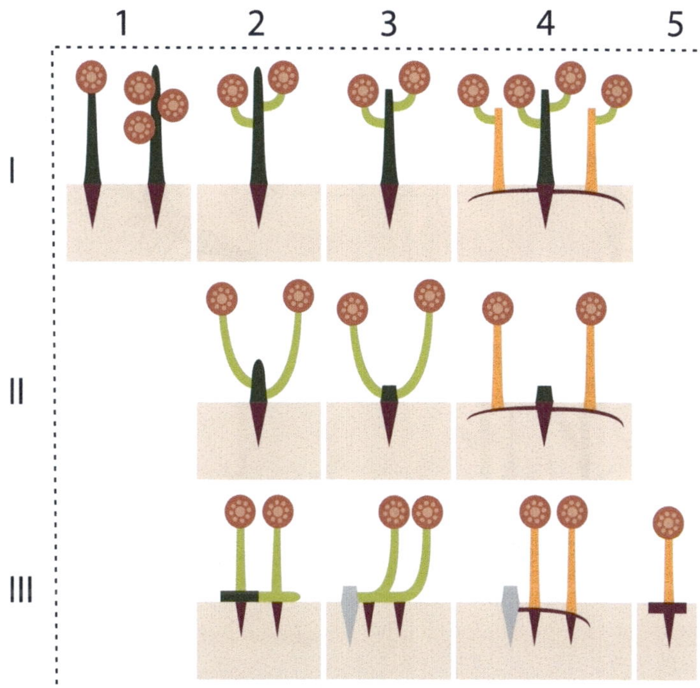

Fig. 3.15 Synthesis of architectural models for herbs (Serebriakova) and woody plants (HOT) with the addition of plants sprouting from roots. 1—the primary shoot is the only shoot throughout the plant's life, 2—the primary shoot grows and is present in the adult body of the plant but axillary shoots are also part of the plant body, 3—the primary shoot is part of the plant body but growth continues through axillary shoots, 4—the primary shoot is part of the plant body but growth continues through adventitious shoots, 5—plants without shoots, I—acrotonic branching (from buds on the upper part of the shoot), II—basitonic branching (from buds on the lower part of the shoot), III—basitonic branching and simultaneously the plant forms adventitious roots that allow the loss of the oldest parts of the plant (and thus splitting of the plant to physically independent fragments, i.e. clonal growth). (According to Klimešová 2018)

Of course, the presented synthetic classification is relatively coarse, and it would be possible to proceed in the opposite way and create as detailed a classification for herbs as for woody plants in the work HOT. It depends on what we want to use such a classification for. For ecological interpretation, coarser categories are advantageous because there are enough species in each category, allowing observations or experiments to be based on a sufficient number of repetitions. For solving evolutionary or developmental questions, it is useful to know the details of the architecture of each species. For annual herbs alone, for example, there is a classification of growth forms based mainly on shoot architecture, which recognizes 20 categories! (Krumbiegel 1998).

Fig. 3.16 Architectural types of herbs. A—herbs with a single shoot throughout life: A1—monocyclic shoot in an annual, A2—polycyclic shoot in a monocarpic perennial (biennial); B—perennial herb with monopodial branching: B1—without a leaf rosette, B2—with a leaf rosette; C—perennial herb with sympodial branching: C1—non-rosette monocyclic shoots, C2—(semi)rosette polycyclic shoots; D—perennial herb with adventitious sprouting from roots: D1—non-rosette monocyclic shoots, D2—(semi)rosette polycyclic shoots. Dead parts of the plant are shown in gray. (According to Klimešová 2018)

3.3.3 Architectural Peculiarities

The fifth type of architecture in our synthetic classification (Fig. 3.15, Table 3.1) is the residual category of architectural peculiarities. We have already mentioned some examples when discussing plant morphology (Fig. 2.2) to show how some plants may lack or have transformed basic organs. Here, I will mention additional examples and emphasize changes in their architecture, specifically in the structure of the primary shoot and root. These are few species, but they are exceedingly interesting not only for their body shape but also for their ecology and evolution. Examples include non-green plants that parasitize the roots of other plants (parasites) directly or through mycorrhiza (mycoheterotrophs). These species have very small dust-like seeds and produce them in large quantities to increase the likelihood of finding a host. The germination of these dust-like seeds, and usually further growth, depends on finding a host, which can be the root of a vascular plant or hypha of a fungus that mediates the connection with the vascular plant. The green primary shoot is not needed by the plant because the key is its connection with the host from which it obtains nutrients (carbon, water, and minerals). Once the

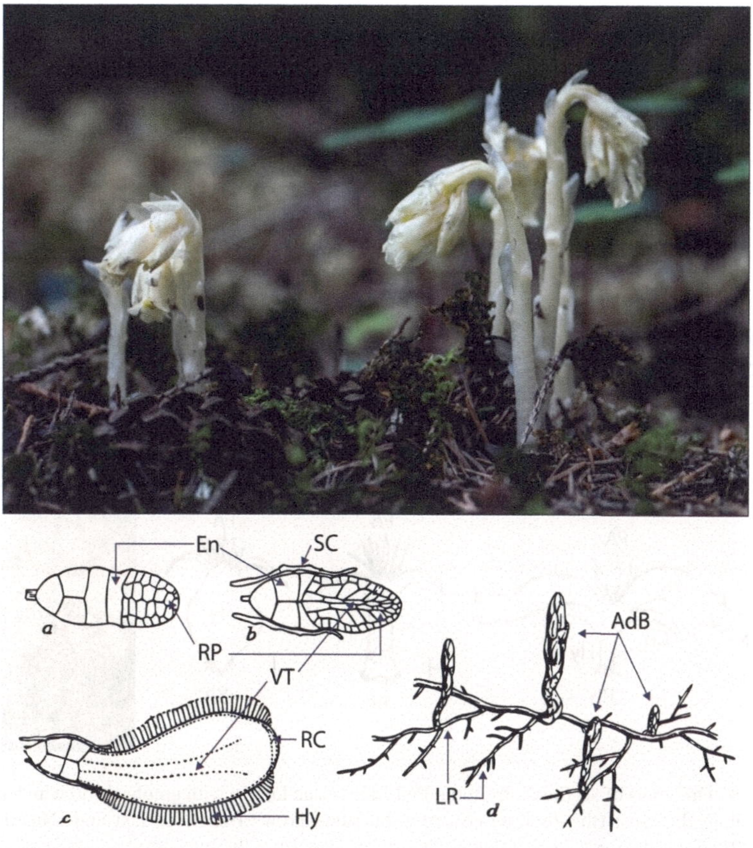

Fig. 3.17 Mycoheterotrophic plant *Monotropa hypopitys* and its germination according to Rauh (Rauh 1937b). Drawings a-c germination phases; (**a**) the stem pole is formed by only a few cells and does not develop further, contrary to the root pole (RP); (**b**) the root pole growths through seed coat (SC) supported by small endosperm (En); (**c**) the root is densely covered with fungal hyphae (Hy), rudimentary root cap (FC), vascular tissue (VT); (**d**) adult plant, which is composed of a branched root system that provides nutrition by parasitizing green plants through mycorrhiza; AdB adventitious buds, LR lateral roots. (Photo Jiří Plekanec)

connection is established, the development of the shoot stops at the hypocotyl stage (for example, species of the genus *Orobanche*) or the shoot completely atrophies, and only the root system is formed (for example in *Monotropa hypopitis*). From these organs, adventitiously flowering shoots then grow, which serve only for seed production and are usually not green (Fig. 3.17).

However, the primary shoot can also be reduced due to a mutation that prevents the plant from growing stems, and the development of the seedling stops at the stage where it has cotyledon leaves, and the apical meristem does not form. Such a mutation would be a dead end if some plants had not learned to bypass it. In nature, we find examples where this handicap has been solved by adventitious sprouting from

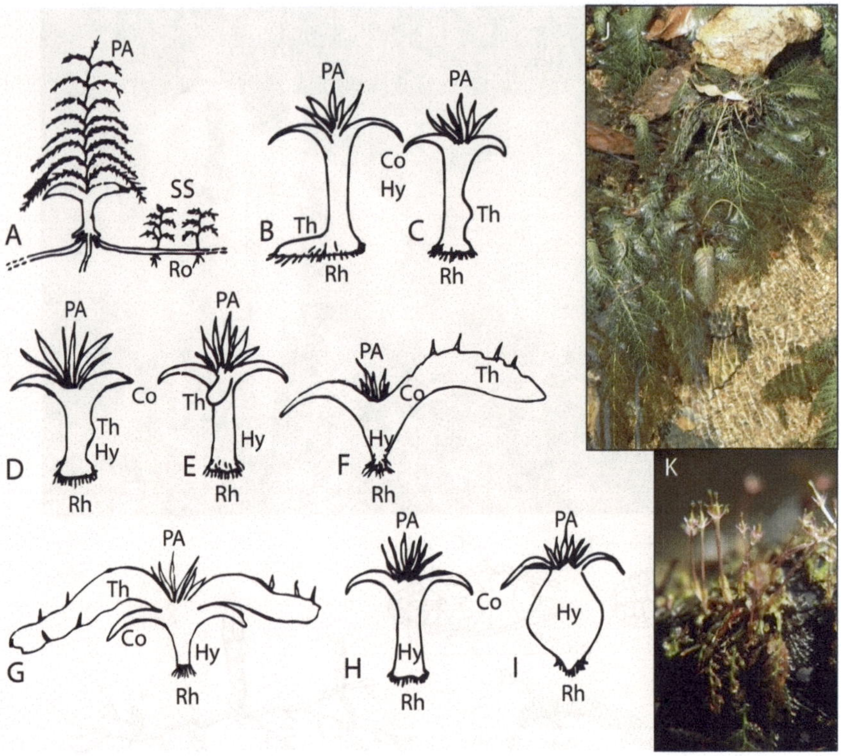

Fig. 3.18 Drawings of seedlings from the Podostemaceae family with emphasis on which part of the seedling the crust (th—thallus) covering the stone grows from (see text and (Nishii et al. 2017)). Type A—does not form a thallus, but roots grow from the hypocotyl and bear secondary shoots (*Indotristicha ramosissima*); Type B—thallus forms from the base of the hypocotyl (*Willisia selaginoides*); Type C—thallus forms from the middle of the hypocotyl (*Zeylanidium lichenoides*); Type D—thallus forms from the upper part of the hypocotyl (*Cladopus hookerianus*); Type E—thallus forms from the same place as the cotyledons grow (*Zeylanidium lichenoides*). Type F—thallus is formed from the cotyledonary leaf (*Polypleurum stylosum*); Type G—thallus forms from the first true leaves (*Polypleurum stylosum*); Type H—the thallus is formed from the epicotyl (the second segment of the primary shoot) (*Hydrobryopsis sessilis*); Type I—the entire seedling contributes to the formation of the thallus. J, K—plants from the genus *Marathrum* from a tropical stream in Belize. (Photo by Petr Macek). Th thallus, Co cotyledonary leaves, PA primary shoot, SS secondary shoot, Hy hypocotyl, Rh rhizoid, Ro root

cotyledon leaves, the hypocotyl, or the root, and moreover, by a rather special way of life that avoids competition with other plants. This has happened at least three times in evolution, in the families Podostemaceae, Gesneriaceae, and Lentibulariaceae.

Plants from the family Podostemaceae are annuals growing in tropical streams and waterfalls (Mohan Ram and Sehgal 1997). The seedling stops its growth at the cotyledon stage, and its root, hypocotyl, or cotyledon leaf begins to overgrow a stone, forming a green crust (thallus) attached by adventitious roots, and from this crust, adventitious inflorescences grow (Fig. 3.18).

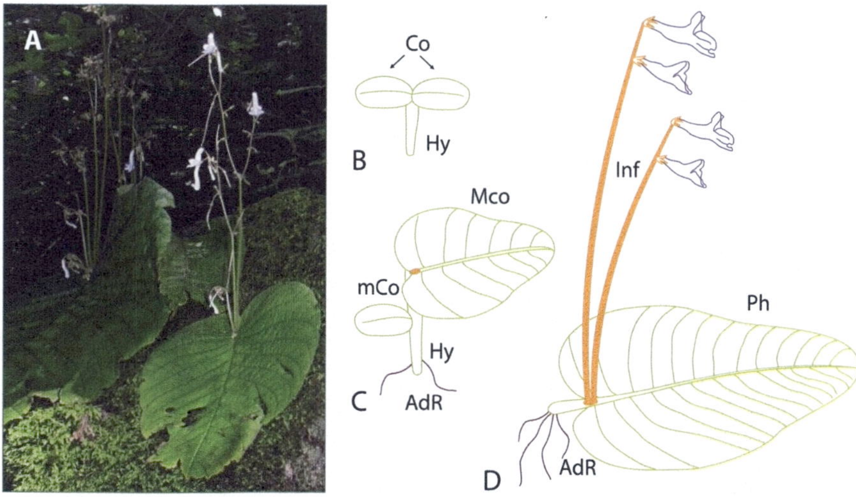

Fig. 3.19 Ontogenetic development of the single-leaf species *Streptocarpus wendlandii*. (According to Nishi et al. 2017). (**A**) overall appearance of the plant; (**B**) seedling; (**C**) juvenile plant, where one cotyledon leaf withers and the other continues to grow; (**D**) flowering plant composed of one leaf, inflorescence, and adventitious roots. Hy hypocotyl, Co cotyledonary leaves, mCo micro-cotyledonary leaf, Mco macro-cotyledonary leaf, AdR adventitious roots, Ph leaf developed from Mco, Inf inflorescence formed on midrib of Ph

Plants from the genus *Streptocarpus* (from the family Gesneriaceae) grow in tropical rainforests mostly as normal plants, but one small group stops growing at the seedling stage (Nishii et al. 2017). One cotyledon leaf falls off, and the other continues to grow until it becomes a large leaf, from which inflorescences grow adventitiously (Fig. 3.19). After flowering, the plant dies. Some species from this large cotyledon leaf produce additional leaves besides inflorescences, and such plants are then perennial and often cultivated as houseplants. Both forms grow as epiphytes, and it is interesting that their relatives, which have a normal body structure, are African violets (*Saintpaulia*), known for readily regenerating using adventitious buds from detached leaves.

In contrast to the aforementioned exotic peculiarities, aquatic carnivorous plants of the genus *Utricularia* from the family Lentibulariaceae can also be found in temperate regions. in nutrient-poor waters. Their development also stops at the seedling stage, where its apical meristem produces a tuft of leaves and does not continue to grow. Fortunately, the plants can transform a leaf into a shoot by beginning to grow indefinitely and having a growth apex that can flower. They manage without roots but have two types of shoots/leaves (Arber 1920): one grows in moist moss and captures prey, the other floats in water or is above water (Fig. 3.20).

Fig. 3.20 Aquatic carnivorous plant bladderwort (*Utricularia*, photo of detail by Jana Navrátilová) with a flowering shoot and a shoot equipped with traps for catching prey. (See detail)

3.3.4 Function of Architectural Models

The first architectural models constructed by Werner Rauh are over 80 years old, but they have only been used to describe the diversity of plant growth and not to address ecological questions. The reason may be that determining them requires morphological knowledge, and to utilize them, we need ecological questions that could be addressed through them. In this regard, architectural models are somewhat similar to classifications of growth forms, with the difference being that they are defined by clear rules.

The fact that architectural models have not yet been used to address ecological questions does not mean that plant architecture, i.e., understanding how the plant body is constructed from repeating shoots, is not useful for ecology. On the contrary, architecture provides tools to describe plant traits that have different ecological significance from the more commonly studied basic plant organ types. For example, vegetative regeneration, clonal growth, or population dynamics are much better understood when viewed through the prism of plant architecture (see relevant chapters).

An interesting example of using the principles of describing architectural models to solve an ecological question is the work on the impact of branching on browsing of woody plants in the South African savanna. Researchers recorded branching parameters on woody plants (where on the parent stem the offspring branch grows, whether at the base, in the middle, at the top, or if it is a continuation of the growth apex), the direction in which the branch grows (horizontally, vertically, or mixed), the shape and length of internodes, the frequency of branching, how many repeating modules are formed, whether thorns are formed, and how hard the wood is. Based on these parameters, the authors constructed an index of "caginess," which was supposed to express how the plant defends against browsing. This index was then related to the degree of leaf browsing by goats of a given species, expressing how many leaves goats are able to graze on the crown surface compared to browsing inside the crown. Woody plants that were densely branched and had a large number of thorns were better able to reduce browsing by goats inside the crown compared to sparsely branched woody plants (Charles-Dominique et al. 2017). The "caginess" index proved to be a suitable parameter of woody plant architecture that evolved as a defense against large herbivores.

Another useful concept of plant architecture is the description of reiteration, i.e., the repetition of the model due to external influences, which can be used in assessing the impact of the environment and disturbance on plant growth. We used it in our work when evaluating the impact of plant injury on seed production (Bartušková and Klimešová 2010). *Rorippa palustris*, which grew undisturbed from seed, had much lower seed production than plants regenerated from roots. This effect was due to the basic architecture being repeated multiple times in disturbed plants, resulting in more branches, flowers, and seeds (see Fig. 3.27 in the chapter on the bud bank).

Recording reiteration on belowground organs is not as easy as on above-ground organs, yet it can help solve ecological questions. Rhizomes form a large part of the biomass of herbaceous communities, thus providing plants with a large bud bank for survival of disturbances (see the chapter on the bud bank) and also protecting the soil from erosion. With increasing intensity of grassland management (more frequent mowing or grazing) causes the biomass of rhizomes to decrease (Ottaviani et al. 2021), and the community becomes susceptible to degradation. To understand the mechanism of this reduced investment in belowground regenerative organs in mowed meadows, it is useful to know how these organs respond to disturbance. Mowing during the production of new belowground rhizome growth causes the new rhizome, instead of reaching the length typical for the species, to stop its horizontal growth after mowing (Klimeš 1999). Its apical meristem does not wait until the next year to produce shoots but grows immediately after mowing. The reason is the disruption of apical dominance and the illumination of the stand. Thus, a rhizomatous plant in a mowed meadow usually has a rhizome system with much shorter annual increments compared to an un-mowed stand. In contrast, plants without rhizomes, such as tufted grasses, grow the same on both meadows (Fig. 3.21). Increasing the frequency of mowing thus leads to a reduction in rhizome biomass.

3.4 Bud Bank

3.4.1 Honeycomb Effect

The term bud bank was introduced by John Harper (Box 24) in his textbook on population biology from 1977 (Harper 1977). He briefly presented it as an analogy to the seed bank, which is very often studied, and emphasized that just as the seed

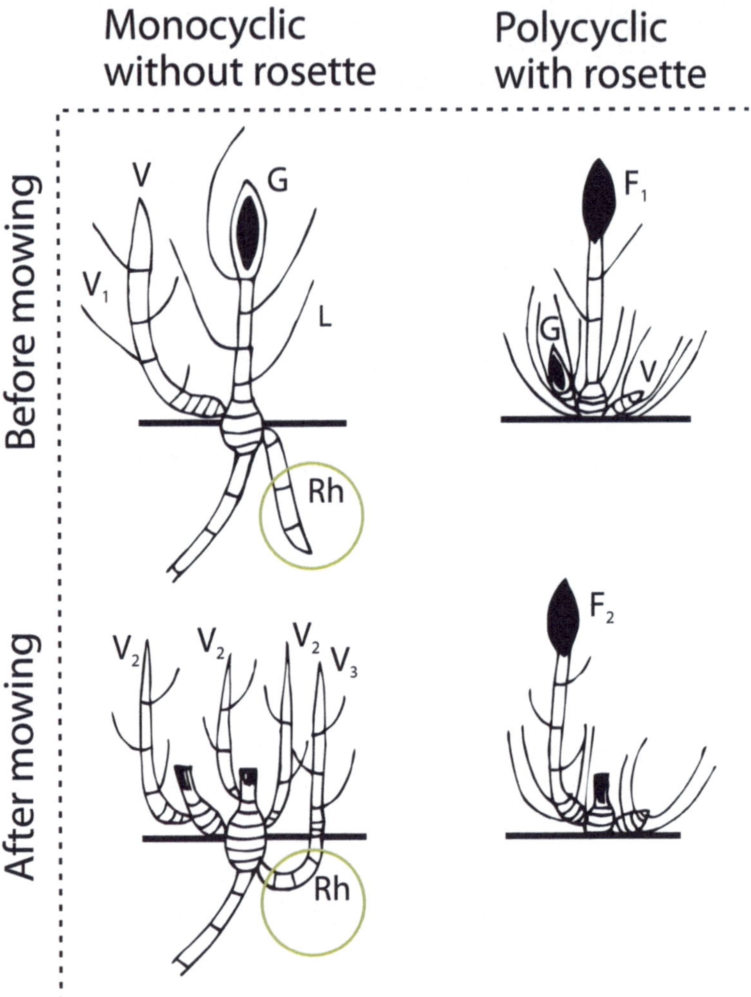

Fig. 3.21 Reaction of rhizomatous and tufted grass to mowing. Rhizomatous grass (left) often also has monocyclic non-rosette shoots, which are removed by mowing (V_1) and new ones grow from the base of the removed shoots (V_2) or from the tip of the rhizome, which instead of continuing belowground growth, grows above-ground (Rh). Tufted grass (right) often has polycyclic shoots with a rosette. Flowering shoots are removed by mowing (F_1), but vegetative rosettes lose only leaves, not the growth apex, and can continue to grow and flower even after mowing (F_2). V_1 and V_2 vegetative shoot, V vegetative growth apex, G generative growth apex, ready to flower, F_1 and F_2 flowering growth apex, Rh rhizome, root system not shown

bank serves plants for growth after disturbance, the bud bank on belowground plant organs, such as rhizomes, bulbs, and tubers, can serve similarly. Although he excluded buds involved in regular seasonal growth, followers pointed out that we usually lack a practical way to distinguish these two types of buds. Buds are ready for use after seasonal dormancy, but when a disturbance occurs that destroys part of the biomass, these buds are used for vegetative regeneration. So, which buds should be included in the bud bank? All on the plant (Klimešová and Klimeš 2007) or only those on specialized regenerative organs protected from disturbance either by their belowground location or by a protective thick bark? (Ott et al. 2019) And what about buds that form on the plant only after injury, as is often the case with adventitious buds on roots or the hypocotyl?

The fact that the entire plant has a modular structure and cannot grow otherwise than by producing new modules and associated buds makes understanding plant adaptation to disturbance difficult. It is something similar to beeswax constructions. Bees build a honeycomb composed on both sides of hexagonal cells inclined so that honey does not spill out. When bees want to build a simple wall in their dwelling, they use exactly the same building elements as when constructing a honeycomb for offspring or storing honey, and the wall has hexagonal cells, even though they are useless there. Similarly, it is difficult for us to decide whether the reason why a plant produced a long horizontal stem (e.g., rhizome) was to spread over a large distance or to produce a larger number of buds.

With mild disturbance, almost all plants can cope thanks to ubiquitous buds (Fig. 3.22). Only repeated disturbance, which destroys all above-ground parts, leads to plants having specialized regenerative organs (see also the chapter on regenerative organs), protecting above-ground buds with bark, or forming adventitious buds after root fragmentation. These buds can thus, for simplicity, be considered part of

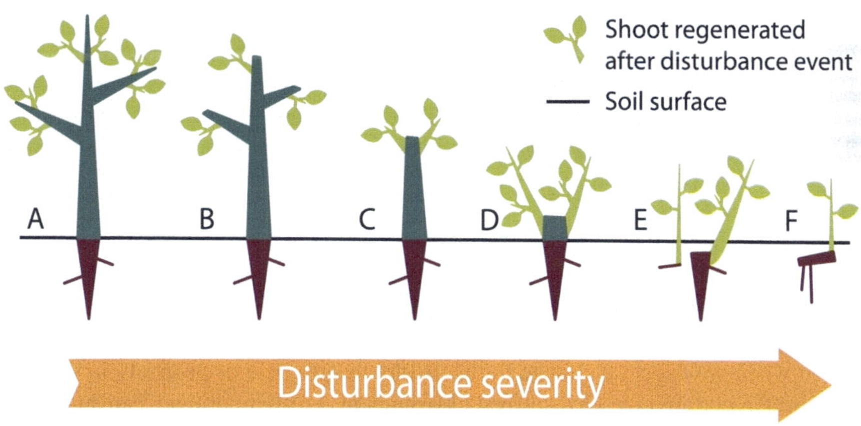

Fig. 3.22 The ability of plants to regenerate after disturbances of varying severity (which we define as the amount of biomass lost). (**A**) only leafy twigs removed, (**B**) lateral branches removed, (**C**) growth apex or tree crown removed, (**D**) most of the above-ground biomass removed, (**E**) all above-ground and belowground stems removed, (**F**) plant fragmented. As the severity of disturbance increases, the number of plants capable of regenerating decreases. While disturbances of strength (**A** and **B**) are managed by all plants, strength (**C** and **D**) are managed by herbs and shrubs, but only a few trees, disturbances (**E** and **F**) are survived by only a few herbaceous species, mostly perennial weeds of arable land. The dark green part of the plant is composed of the stem and the purple part of the root. (According to Klimešová and Klimeš 2003)

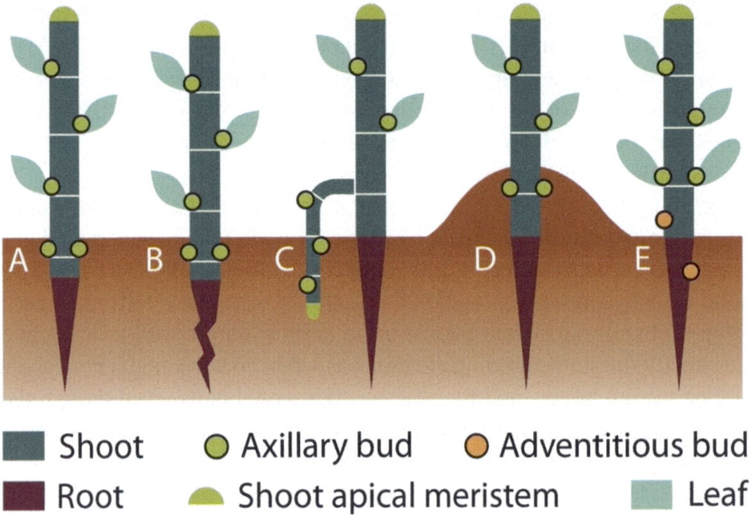

Fig. 3.23 Ways in which the bud bank in juvenile plants can get belowground. (**A**) belowground germination; (**B**) contractile roots; (**C**) positively geotropic stem growth; (**D**) burial by litter or substrate; (**E**) adventitious buds on roots. (According to Ott et al. 2019)

the bud bank because the plant invests energy in them, which could be used for taller growth, competitive ability, or seed production.

Additionally, for plants to store regenerative organs belowground, they must modify growth so that stems, which normally grow upwards during seed germination, are placed belowground or they must be capable of creating buds on roots (Fig. 3.23). If we prevent disturbance from recurring in a community that is adapted to regular disruption, plants with belowground organs are at a disadvantage and are gradually replaced by plants that invest in vertical growth instead of a belowground safety net. When the disturbance returns, these non-adapted plants are budless, thus the vegetation is destroyed.

Testing the hypothesis responsibly, whether a given bud bank is truly an adaptation of the plant to disturbance, is not easy at all, and it is necessary to use various approaches that together can indicate how it is with that bud bank. One possibility is phylogenetic analysis, which examines whether a certain trait appeared during a period when we expect a change in the disturbance regime or precisely in species inhabiting a certain type of disturbed environment. Furthermore, we can study the natural gradient of disturbance frequency or intensity and see if the parameters of the bud bank (for example, the number of buds, depth of burial belowground, or the plant's ability to create adventitious buds) are somehow related. A useful option is an experiment where we manipulate the disturbance and/or the parameters of the bud bank and observe how the plants fare. Since we used these methods in our research, I will elaborate on them a bit more here.

We used phylogenetic analysis to test the question of whether the ability of plants to sprout from roots developed in Central European species to help them survive disturbances (Klimešová et al. 2017). The results suggest that sprouting from roots is a trait that plants can easily gain and lose in evolution. This type of bud bank is more common in plants of heavily disturbed habitats, but the species growing there usually acquired the ability to sprout from roots only after they colonized these habitats.

When we analyzed the size of the bud bank along the disturbance gradient for the Central European flora, we found that the largest bud banks on belowground stem organs are found in plants where severe disturbance is of medium frequency and does not recur too often (once a year) or too rarely (once every hundred years) (Herben et al. 2018). This corresponds to predictions from the literature, which consider investment in a bud bank purposeful when a strong disturbance returns every 5–30 years (Fig. 3.24) (Bellingham and Sparrow 2000). If the disturbance is rarer, even trees that do not invest in a bud bank can complete their life cycle; if it is more frequent, the regenerative organs cannot replenish their reserves (buds and carbohydrates), and the plant is unable to sprout. Here, in places with frequent severe disturbance, short-lived plants dominate, because they manage to complete their life cycle between two disturbances and do not invest in regenerative organs. Or there are plants that form buds only after disturbance and on roots.

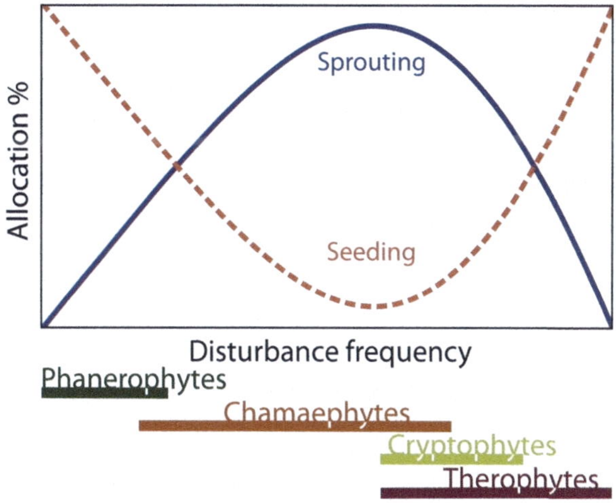

Fig. 3.24 Empirical model predicting investment (proportional allocation of resources available to the plant) into vegetative regeneration versus generative regeneration in plants along a disturbance frequency gradient. In conditions where disturbance occurs once every hundreds of years (left part of the graph), trees (phanerophytes) reproducing by seeds dominate. Where strong disturbance returns between every few decades to a few years (middle part of the graph), shrubs and perennial herbs (chamaephytes and cryptophytes) dominate, regenerating from belowground organs after disturbance. When strong disturbance comes every year or more frequently (the right part of the graph), annuals dominate, regenerating from seeds and additionally, in these conditions, there is a small group of species that form adventitious buds on roots (not shown in the figure). (According to Bellingham and Sparrow 2000)

Both studies thus supported our hypothesis that sprouting from roots is an adaptation to strong disturbance. However, we also attempted experimental testing, which I will present to you in the following chapter.

3.4.2 Experimental Testing of Function

In the experiment, we can easily simulate different types of disturbance; it is more challenging to manipulate the bud bank and decide which parameter of plant fitness to evaluate. Although it seems like a tricky problem, it can be solved with an appropriate selection of model plants, which we attempted in our research. We used sprouting from roots and the hypocotyl in short-lived plants, i.e., annuals and biennials, as a model. The advantage of this model system is that buds form only after disturbance and cost the plant nothing beforehand. Moreover, in short-lived plants, we can determine lifetime fitness as seed production because they reproduce only once in their lifetime.

First, we had to solve the problem of what to compare the disturbed plants and their ability to regenerate and resulting lifetime fitness with. Imagine that in a population of annual plants without a bud bank on belowground organs, there is a mutation that allows individuals carrying the mutation to sprout from the root and complete the life cycle, thus producing offspring and passing on the same ability. We are interested in the conditions under which this ability spreads in the population. The arrangement of the experiment (and especially the definition of control) depends on the level of disturbance we expect. We tested two scenarios (Fig. 3.25): (i) we assume a disturbance that affects only some plants in the population, so we compare disturbed and undisturbed plants; (ii) we assume a disturbance that affects all plants in the population, so we compare disturbed plants with plants that germinated at the time of disturbance from seeds.

The first scenario of comparing disturbed and undisturbed plants of the same age is often used in research on the impact of browsing. A large herbivore finds only some plants in the population, bites them (usually only their top part), and the other plants grow undisturbed. In testing this scenario with short-lived plants sprouting from the root after strong disturbance, we found that if a plant is disturbed early in ontogeny and also early in the season, it usually produces the same or a greater number of seeds than an undisturbed plant. This is because the disturbance removes the dominance of the growth apex, causing the plant to branch more than an undisturbed plant. The result is the same as in classic experiments with plant apex browsing (Huhta et al. 2003), but in our experiment, we removed the entire above-ground biomass of our plants. Because the disturbed plant has a shorter time for seed production, the quality of the seeds (their size and germination) is often lower than the quality of seeds from an undisturbed plant (Martínková et al. 2008). The evolutionary advantage of sprouting from a bud bank and thus the spread of a hypothetical mutation allowing sprouting from roots depends on the timing of disturbance and the selective pressure on seed quality.

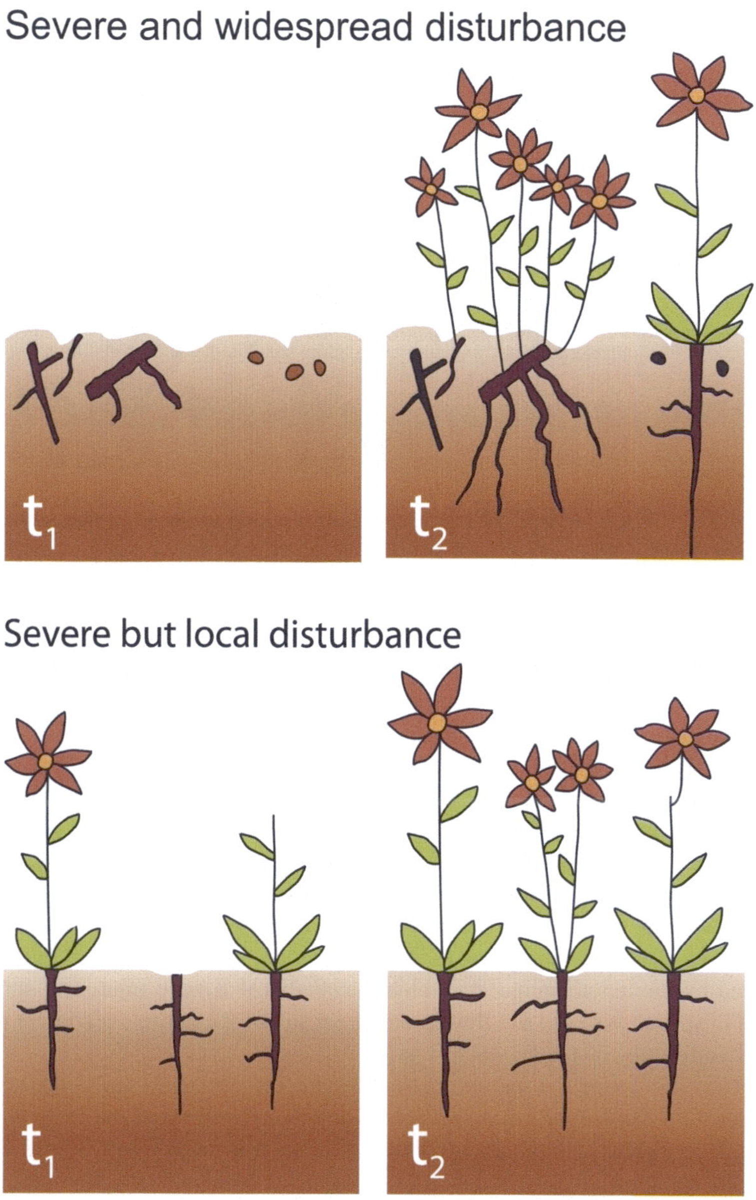

Fig. 3.25 Two scenarios of disturbance in experiments testing the functionality of the bud bank on roots and hypocotyl in short-lived plants (for example *Rorippa palustris*). Upper panels show the severe and widespread disturbance during which all aboveground biomass of a community together with some belowground biomass is removed (t_1). After such widespread disturbance plants that regenerated from root fragments compete with plants regenerated from seeds (t_2). Lower panels show severe but local disturbance during which some plants lost aboveground biomass, and some remain undamaged (t_1). After such disturbance plants that regenerate from root fragments compete with undisturbed plants (t_2). (According to Klimešová and Martínková 2022)

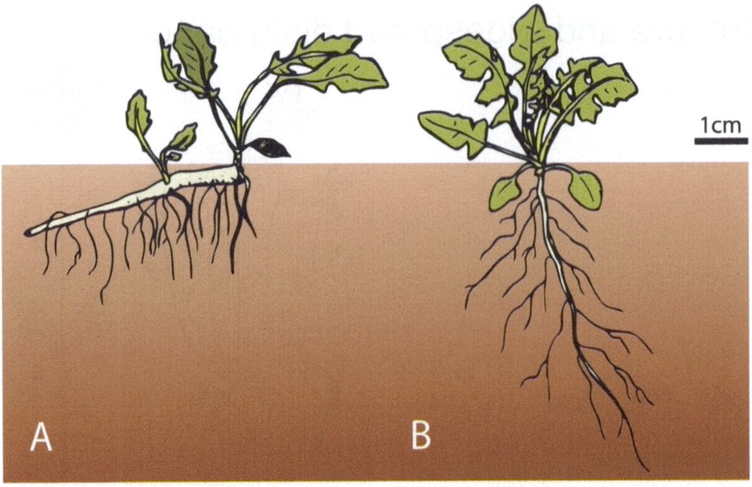

Fig. 3.26 Plants of *Rorippa palustris* regenerating after severe disturbance. (**A**) vegetative regeneration from a root fragment, (**B**) generative regeneration from a seed

In testing the second scenario through an experiment, we found that plants grown from root fragments after simulated strong disturbance at the beginning of the season had the advantage of large carbohydrate reserves in the root fragment and greater branching (due to several adventitious buds that formed on the root; Fig. 3.26). From a seed only one shoot grew while from a root fragment several regenerated shoots emerged and each of them was as large as the single shoot from a plant grown from seed. Therefore the seed production of regenerated plants was several times greater than the seed production of plants germinated at the time of disturbance from seeds (Klimešová et al. 2008) (Fig. 3.27). However, we were not satisfied with this result because it did not capture the variability of conditions that can occur in nature. The flowering of the experimental species, *Rorippa palustris*, is triggered by the lengthening day, which means that if it were injured in summer instead of spring, it would not flower, and the result of the entire experiment could differ significantly.

We therefore tested the second scenario again. This time we established several experimental cohorts of *Rorippa palustris*, in which we allowed the experimental plants to germinate at monthly intervals from April to September. While undisturbed control plants that germinated in spring behaved as annuals, plants that germinated just before the summer solstice mostly behaved as short-lived perennials, i.e., they bore fruit in the first year, survived the winter, and bore fruit again the next year. Plants that germinated after the summer solstice behaved as winter annuals and postponed flowering, triggered by the lengthening day, until the following spring (Klimešová et al. 2007). This variability in the life cycle is caused by the fact that for the germination of *Rorippa palustris* seeds, a temperature fluctuation of about 10 °C between day and night is sufficient, as this signals that the place is not

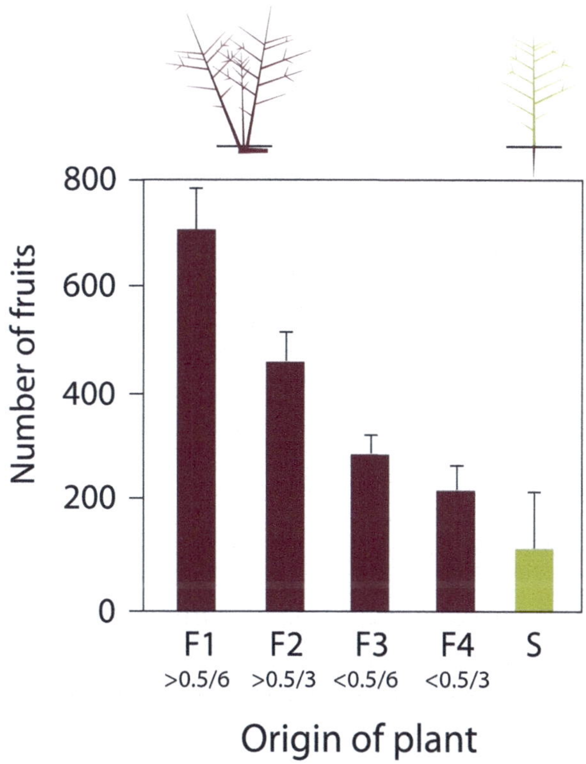

Fig. 3.27 Fruit production in *Rorippa palustris* plants regenerating after disturbance from root fragments (F1—F4) and from seeds (S). F1—root fragment thicker than 0.5 cm and 6 cm long, F2—root fragment thicker than 0.5 cm and 3 cm long, F3—root fragment thinner than 0.5 cm and 6 cm long, F4—root fragment thinner than 0.5 cm and 3 cm long. Above the graph, the branching of experimental plants is schematically shown. (According to Klimešová et al. 2008)

overgrown with vegetation and will not have to struggle with competition. However, the flowering of the plant is controlled by the length of the day. The plant blooms when exposed to lengthening days before the summer solstice and flowers independently of the plant's size. Vegetative regeneration of plants after disturbance in our experiment was similarly diverse and dependent on timing as in the control for the same cohorts of the plant. The best regeneration from roots occurred in plants disturbed at the rosette stage, especially if they were disturbed early in spring.

Based on the results of this experiment, we modeled how populations that germinated and were disturbed at different times would behave in the long term (Sosnová et al. 2014). Spring disturbance of plants was the only timing that led to population growth if the success (intensity in Fig. 3.28) of root sprouting increased. Otherwise, the rate of root sprouting did not matter, and the population grew due to regeneration from seeds, especially if disturbed in the second part of summer or autumn (Fig. 3.28). The bud bank on the roots of this species with very variable ontogeny is advantageous when disturbance is frequent and its timing remains the same, as is the case on arable land. In nature, where plant disturbance is less predictable, the plant manages with a variable life cycle and can regenerate from seeds at any time.

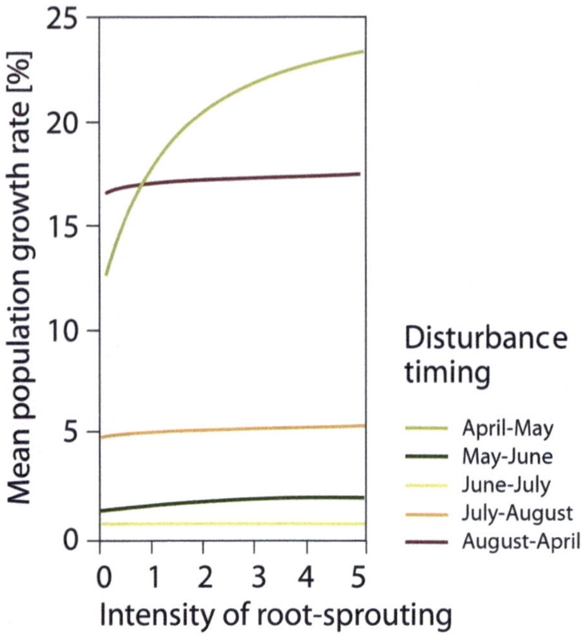

Fig. 3.28 Results of the population growth model of *Rorippa palustris* with different timing of severe disturbance, which affects the population once a year, and the population has varying abilities to sprout from roots (x-axis). The intensity of root sprouting affects mean population growth rate only when the population is disturbed in spring (April–May). Spring timing along with disturbance during autumn and winter leads to the highest population growth, but this is due to regeneration from seeds. (According to Sosnová et al. 2014)

3.4.3 Bud Bank in Seasonal and Disturbed Communities

The bud bank has not yet been studied in all disturbed communities. We have the most data for prairies and steppes, less for tropical grasslands. In shrub communities exposed to recurring fires (such as hard-leaved vegetation on Fig. 1.1), it is traditionally examined only whether shrubs survive the fire or regenerate from seeds, and quantification of the bud bank for these communities is lacking. Similarly, on arable land, studies prevail that distinguish from which special organs plants regenerate but do not address the quantification of the bud bank (see the chapter on belowground organs).

In plant communities such as steppes or meadows, most plants that grow in spring or after a dry period grow from the bud bank and not from seeds (Benson and Hartnett 2006). Since the bud bank is usually associated with a reserve of carbohydrates in belowground organs, small seeds in such vegetation have little chance to establish and wait for small-scale disturbance that suppresses the growth of surrounding perennial plants. The bud bank not only allows seasonal growth and growth after disturbance, but it is also resistant to weather fluctuations that can cause the inability of plants to germinate from seeds. This is especially important at the interface of grassland and semi-desert ecosystems. Grasslands composed of perennial species have much more stable primary biomass production than annuals in semi-deserts, which experience large fluctuations depending on current weather, which triggers the germination or not (Dalgleish and Hartnett 2006).

In the American prairies and the steppes of Inner Mongolia (Qian et al. 2017), the boundary between steppe and semi-desert in terms of the bud bank has been described, so we know that at this interface, there are only about 100 buds per square meter. In the moist parts of steppes and prairies, the number of buds per square meter is several thousand. Extreme dry events and excessive grazing can further reduce the bud bank in steppes (see also the example of rhizome growth after disturbance, Fig. 3.21), increasing the risk of soil erosion and the danger of replacing perennial plants with annuals that do not protect the soil against wind erosion at all. Both regions have their experiences with this. In the American prairies, only fragments of the original vegetation remain, most prairies were plowed into agricultural land. During the dry period in the 1930s, this led to catastrophic wind erosion. In the Mongolian steppes in northern China, due to excessive grazing intensity, some grasslands have turned into annual communities, and in winter, when the soil is not covered by vegetation and mostly not even by snow due to the small amount of precipitation, significant wind erosion occurs.

The role prairie fires played before the colonization of America and the extermination of bison is unclear, but it is assumed that they were frequent and of low intensity because there was little material available to burn—bison grazed everything. Today, the area is agriculturally used as fields and pastures, but the intensity of pasture use is decreasing. Since the pastures no longer host as much livestock as before, woody plants are beginning to spread. The accumulation of fuel is ideal for the occurrence of uncontrolled fires, which are all the worse because human dwellings are scattered in the landscape and can be easily threatened by fire (Pyne 2017). The largest fragment of prairie that was never plowed, Konza Prairie, located in the Flint Hills of Kansas, is a large experiment where the effects of different fire regimes, drought, and grazing on plants and their bud bank are studied. Short-term fluctuations are well tolerated by plants, and the bud bank does not change because of them (Vanderweide and Hartnett 2015). However, if the frequency of fires is too low, grazing alone will not prevent woody plants, which are not adapted to fire, from spreading into the prairie.

Tropical grasslands in Africa, South America, and Australia are not entirely grasslands because they host a large number of woody plants, mostly shrubs. The main type of disturbance here is browsing by large herbivores (mainly in Africa) and regularly returning fires. Woody plants, like herbs, have belowground organs with a bud bank from which they can regenerate (Pausas et al. 2018). They often also have thick bark on above-ground stems, which protects the buds from fire. Structures that allow regeneration after disturbance are costly, and plants that produce them grow slowly and are small in stature. If grazing and fire are excluded in Brazilian tropical grasslands called Cerrados, herbs and small shrubs that invested in building belowground storage organs with a bud bank are replaced by tall shrubs and trees that invest in above-ground growth and competitive ability (Fidelis et al. 2014). When such a community is affected by fire, these competitively strong woody plants are unable to regenerate from the bud bank and perish—the community no longer has the bud and carbohydrate reserves needed for regeneration (see also regenerative organs).

3.4.4 What Limits Regeneration from the Bud Bank

A plant produces new buds simultaneously as it produces new stems, but this does not mean that the number of buds on the plant constantly increases. In North American grasses, it has been found that the lifespan of a bud is about 2 years and can thus live a shorter time than the stem (rhizome) on which it is located. In secondarily thickening stems and roots, dormant buds must grow at least a little each year to remain on the surface of the organ and not be overgrown by wood. They wait for their opportunity to grow for decades. The timing of stem natality and mortality can cause fluctuations in the bud bank, making the plant more vulnerable to disturbance at certain times of the year—this is used, for example, in the fight against perennial weeds. We observed this effect in our experiment with *Rorippa palustris*, but other plants sprouting from roots also have periods when they are more affected by disturbance and when farmers are advised to perform weeding interventions (Verwijst et al. 2018).

The insufficient size of the bud bank can also be the result of repeated disturbance when the bud bank is depleted. This can mainly occur with the bud bank on stems, because the number of buds there is determined by the modular structure. The bud bank on roots is much more difficult to quantify; buds may be established on roots before disturbance but may not be visible on the root surface and can also be established only after disturbance (so-called regenerative buds). In woody plants, it is assumed that regenerative buds, established on roots only after disturbance, cannot serve for sprouting because the plant is unable to connect these buds to the vascular bundles. This does not seem to apply to herbs, where soft tissues do not prevent the connection of adventitious buds with the vascular system and can serve for regeneration.

Regeneration of the plant may not only be threatened by a depleted bud bank, but carbohydrate reserves in the regenerative organ may also be depleted, or there may be a lack of water and nutrients in the soil (Wise and Abrahamson 2008). All these are resources necessary for building new biomass, and their lack can cause poor plant regeneration, for example, on intensively used pastures.

3.5 Clonality

3.5.1 Clonal Traits

Clonal growth arises if a plant originating from a seed, a genet, has the ability to break down into multiple physically independent rooting units, ramets. As pointed out by Groff and Kaplan (see above), this can happen when the plant is capable of producing adventitious roots from stems, adventitious shoots from roots, or both simultaneously, resulting in more rooting units (Fig. 3.14). However, there are significant differences among clonal plants, and it is therefore important to address these differences and determine if they have any ecological significance.

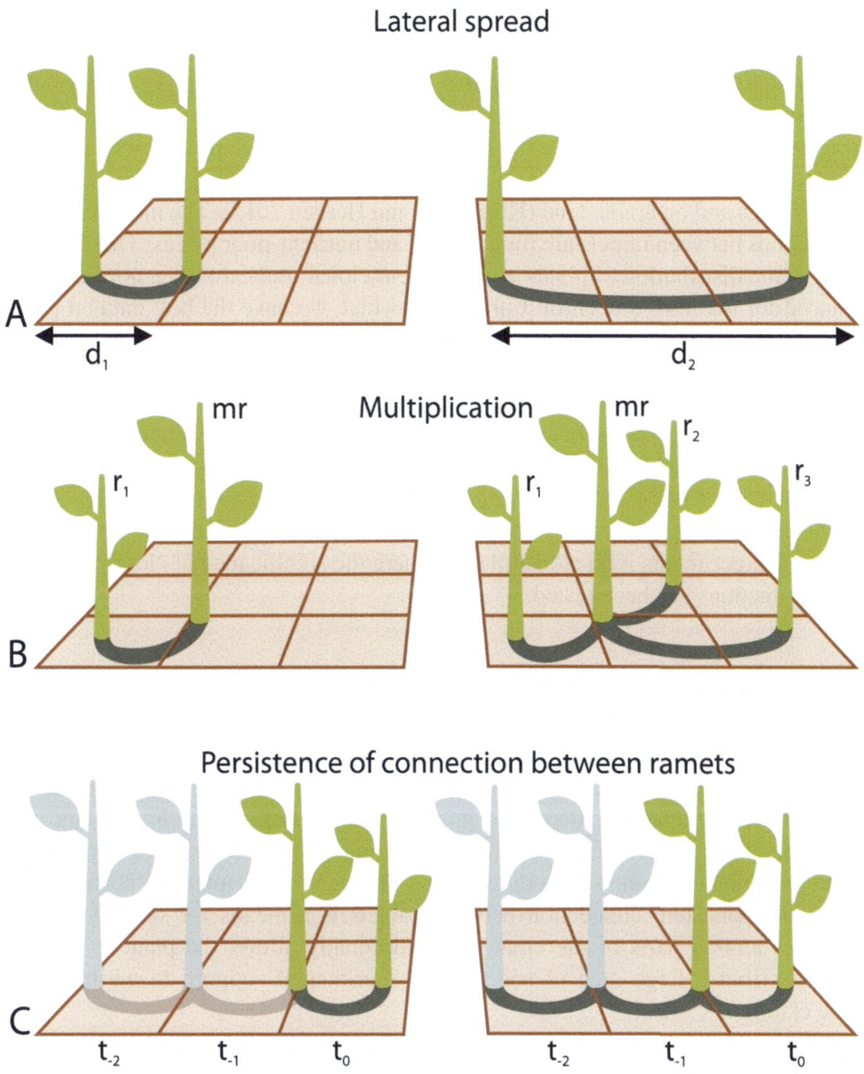

Fig. 3.29 Clonal growth functional traits. Lateral spread—the distance between the parent and offspring shoot (ramet) grown in 1 year, Multiplication—the number of offspring shoots per parent shoot per year, Persistence of connection—the number of years the offspring shoots remain connected. Schematics in the left column indicate small values of the trait, in the right column are large values. Green indicates living parts of the clone, gray, dead parts

We can define clonal traits across the morphological categories we introduced in the chapter on comparative morphology. The most important can be identified as lateral spread, multiplication, and persistence of connections between ramets (Fig. 3.29). These traits are partially correlated with each other. Plants that spread laterally over long distances also usually produce many offspring ramets, and the connection of these ramets is short-lived (see (Bartušková et al. 2022), while (Chelli et al. 2024) shows that persistence is independent on lateral spread). Similarly, as morphologically defined organs of clonal growth (such as rhizomes or runners) are responding in their occurrence along environmental gradients, clonal traits are even more sensitive. Lateral spread, for example, increases from dry or cold areas to moist and warm conditions, while the number of clonal offspring is high in aquatic communities and on arable land (Klimešová and Herben 2015). The most persistent connections between ramets are found in dry and nutrient-poor places. These differences in the distribution of plants with different clonal traits are often not very pronounced for the flora of Central Europe (for which we have the best data on plant clonality), as it does not have truly dry or cold environments here. However, when we look into other areas with more extreme climates, these trends are confirmed. While in the dry grasslands in Central Europe, plants with short lateral increments prevail, but a few species with long increments also occur, in the Mongolian steppes, where dryness is much more pronounced, no plants with long rhizomes are found at the driest end of the gradient (Qian et al. 2017). How to interpret the behavior of clonal plants along gradients? To answer this question, we must delve into the results of experiments with clonal plants, where the significance of clonal traits for certain functions has been tested.

3.5.2 Functions of Clonal Traits

Clonal growth is defined as the production of potentially independent individuals, ramets, but its significance is not only in plant demography (this will be discussed in the following chapter): thanks to morphological plasticity, a clonal plant can place ramets in favorable, for example, nutrient-rich environments; support of young establishing ramets in a dense stand allows a clonal plant to occupy new space; the persistent connection between ramets is a means for sharing limited resources among parts of the clone and additionally allows the plant to harvest nutrients from a large area of soil. What experimental support do we have for these claims?

Resource searching through clonal growth, known as foraging, has been studied in many pot experiments. Clonal plants are usually left to grow in a heterogeneous soil environment, where areas of nutrient-rich and nutrient-poor soil alternate, and in control homogeneous conditions; in both variants, the overall availability of nutrients is the same. After some time, the plants are harvested, and it is assessed whether the distribution of biomass of stems, roots, and the placement of offspring ramets relative to the heterogeneity of the substrate is random. After conducting dozens of similar experiments, we can say that plants usually preferentially place roots in nutrient-rich soil, but the placement of ramets is rather random. This applies to plants with stolons, rhizomes, or roots as the organ of clonal growth (Martínková et al. 2018) (Fig. 3.30). Targeted resource searching through clonal growth is therefore not a universal property of clonal plants, although it is occasionally recorded in some combinations of soil heterogeneity mosaic size and plant size. On the

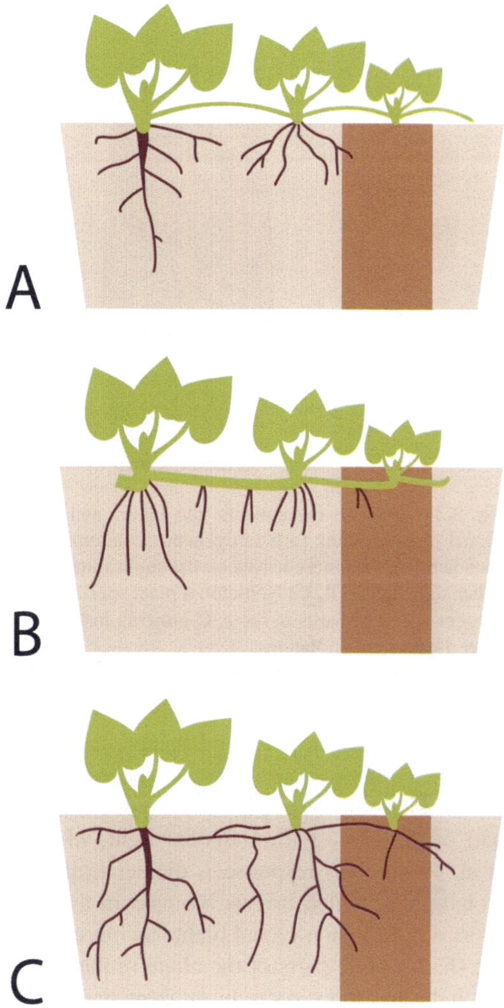

Fig. 3.30 Experiments with heterogeneous nutrient distribution in the soil did not prove that clonal plants preferentially place ramets in nutrient-rich areas. This hypothesis was rejected for plants with stolons (**A**), rhizomes (**B**), and plants sprouting from roots (**C**)

contrary, in experiments examining heterogeneous environments above-ground, unshaded and shaded places, researchers studying clonal plants were more successful: preferential placement of offspring ramets into unshaded places by clonal plants spreading with stolons (but not rhizomes) was confirmed (Xie et al. 2014).

Another frequently tested function of clonality in experimental conditions is the sharing of resources between connected parts of the clone in a heterogeneous environment, i.e., division of labor in clonal plants. Researchers test the hypothesis that the part of the clone growing in a place where a limiting resource is easily available specializes in acquiring it and shares it with the part of the clone growing in conditions lacking this resource. Usually, a heterogeneous soil environment in terms of nutrients is used for testing, combined with shading of the above-ground parts of the

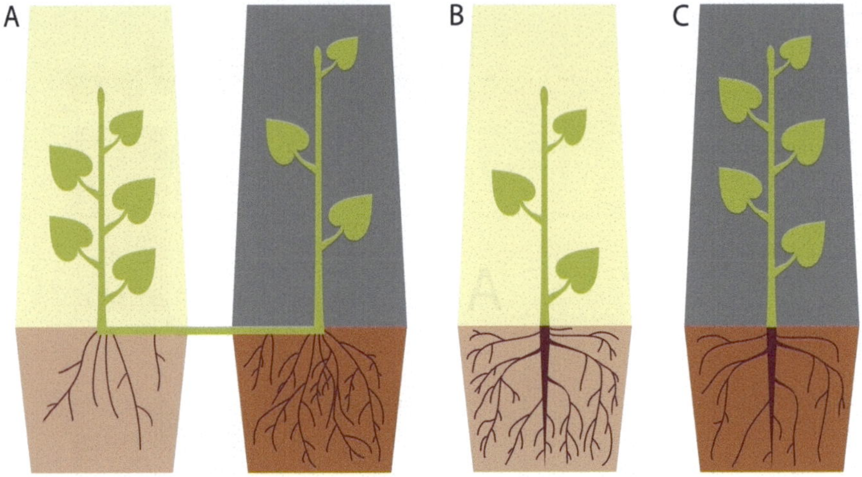

Fig. 3.31 Relative investments into leaves and roots depending on resource availability. (**A**) Clonal plant growing in heterogeneous conditions; its ramets specialize in the resource that is abundant, i.e., has more leaves on the sunlit ramet (left) and more roots on the ramet growing in richer soil (right). (**B**, **C**) Non-clonal plant specializes in the resource that is scarce (**B** growing in nutrient scarcity invests in roots, **C** growing in light scarcity invests in leaves). Shaded plants are taller due to stem etiolation

plant. Part of the clone is thus exposed to poor soil but full sunlight, and part of the clone to nutrient-rich soil and shade.

Clonal plants in the experiments responded according to the hypothesis and invested in roots in the part of the clone with abundant nutrients and in leaves in the part of the clone with abundant light, while sharing limiting resources (Stueffer et al. 1996). This response is diametrically different from the response of a non-clonal plant, which invests biomass in the organ acquiring the resource that is in short supply. A non-clonal plant has more fine roots in poor soil and more leaves in the shade, while parts of the clone in these environments will have more leaves in the light and more roots in nutrient-rich soil (Fig. 3.31). Most experiments were again based on examining plants that are clonal due to stems and only a few on plants that are clonal due to roots. It seems that unlike stem-clonal plants, root-clonal plants exchange resources only minimally, and division of labor among them does not exist (Klimeš and Klimešová 1999).

Division of labor does not only occur in a clone growing in heterogeneous conditions but also when parts of the clone differ in their ability to acquire resources, which is the case of parental versus offspring ramets that have not yet developed a root system or leaves. This function, where the older part of the clone supports the younger part, might more appropriately be likened to parental care rather than division of labor (Stuefer 1998). Some ramets break off from the parent plant almost immediately, and the parent invests only slightly more into them than into seeds, thus facilitating their establishment in the community only marginally. Other ramets are supported during the first year of life, when they develop their own leaves and roots, while some ramets remain connected to the rest of the clone for decades. Parental support for establishing ramets gives clonal plants an advantage in establishing themselves in dense community cover, where seed regeneration is very difficult. It is no wonder that in grasslands up to 98% of shoots are ramets and only about 2% are plants established from seeds (Benson and Hartnett 2006).

As we have already mentioned when describing experiments with the division of labor among parts of the clone, it is advantageous for the clone to be connected precisely because it allows for resource sharing. However, in nature, we rarely have such contrasting environments as we use in experiments; substrate heterogeneity and shading are usually milder and variable over time. In environments with low nutrient availability, however, the mere fact that the clone roots over a large area of soil can be important for plant growth. It allows it to harvest nutrients from a larger area than its above-ground biomass covers. This harvesting from a large area is likely at the expense of deep root growth, as shown by preliminary analyses. There are still few specific studies on this topic, especially for adult plants, but there are experiments with seedlings of clonal and non-clonal plants and their root growth strategies.

The root system of plants that will be clonal in adulthood differs from the root system of plants that will not be clonal. Non-clonal plants will have a main root throughout their life, so even as young plants, they invest in its growth, root deeper, and have more branched roots than clonal plants (Šmilauerová and Šmilauer 2007). Clonal plants will have to form a new root system on new rhizome growths, and the roots on old parts of the rhizome will die along with the rhizome—even young plants have shorter and less branched roots than non-clonal plants. Poorly branched roots are typical for fast-growing species that do not bother with exploring already occupied soil and continuously grow into new areas. These morphological differences in the root system of clonal and non-clonal plants correspond to the results of an experiment (Weiser et al. 2016), which examined how clonal and non-clonal plants respond with root growth to heterogeneity in nutrient availability in the soil. Again, these were young plants that had not yet become clonal. The plants were placed in the middle of a pot, with a solution with a high nutrient content dripped on one side of the pot, water on the other side, and a control where both sides of the pot received the same average watering (Fig. 3.32). Clonal plants, unlike non-clonal ones, did not bother with accumulating roots only in the half of the pot with rich watering; they simply grew everywhere equally. Growth directed into the rich part of the pot apparently does not pay off for them because the roots are short-lived, and investing in signaling and asymmetric growth is not worthwhile, as they can exchange resources within the clone anyway. Sooner or later, the clonal plant will "encounter" the rich part of the soil during its horizontal growth.

3.5.3 Phalanx and Guerrilla

A simple and very popular classification of clonal plants divides them into two groups: those that form a compact clone, so offspring ramets are very likely to meet

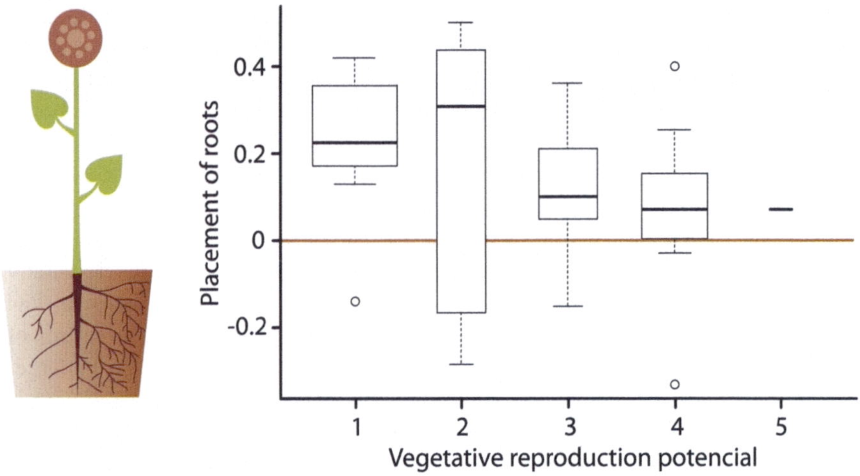

Fig. 3.32 Experiment with seedlings growing in a heterogeneous environment. Seedlings that will later in life become clonal plants did not preferentially place roots in richer part of the substrate, especially when they had a high potential for vegetative reproduction. The difference in root placement in heterogeneous versus homogeneous conditions is shown on the y-axis (0 = control), potential for vegetative reproduction on the x-axis from small lateral spread and production of offspring ramets (1) to large values of these properties (5). (According to Weiser et al. 2016)

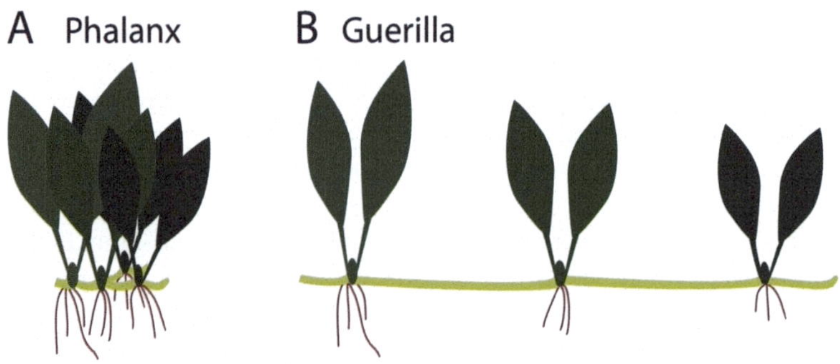

Fig. 3.33 Growth forms of clonal plants. (**A**) Phalanx: ramets form a compact stand and thus mainly neighbor individuals of the same species (e.g., *Allium ursinum*), (**B**) Guerilla: ramets do not form a compact stand and mainly neighbor individuals of other species (e.g., *Fragaria vesca*)

and compete for resources mainly with ramets of the same clone (Phalanx), and those that form a more dispersed clone, where individual ramets are more likely to encounter other plants than members of their own clone (Guerrilla) (Fig. 3.33). Phalanx denotes a compact formation of troops, while Guerrilla refers to the guerrilla warfare of solitary fighters. These terms are therefore more military than botanical, but the authors likely wanted to suggest that among individuals, and consequently species within a community, there is more of a struggle than coexistence. These terms first appeared in 1981 in an article by Leslie Lovett-Doust (1981) about the creeping buttercup and were used to describe intraspecific variability. Today, they are used not only for intraspecific variability but also to describe interspecific differences in growth strategies.

Researchers have focused, for example, on the growth of plants with these contrasting strategies in experiments. It has been confirmed that plants capable of placing offspring ramets both far and near from the parent ramet invest more in the Phalanx strategy than in the Guerrilla strategy in conditions of high nutrient availability. This means they specialize in utilizing resources at a given location. Conversely, plants growing in poor conditions invest more in the Guerrilla strategy and try to leave the inhospitable place (Ye et al. 2006). Similar behavior of clonal plants has been confirmed in studies using in silico experiments, i.e., in computer models.

There are also studies that ask how the representation of Phalanx and Guerrilla strategies changes in clonal plants along environmental. When we look closely at these studies, we find that the trait used by the authors to distinguish strategies is not whether the neighbors of the ramets are plants of the same species or another species, but whether the clonal plant has long or short lateral spread (e.g., rhizome increment per year) (Song and Dong 2002). However, this is not the best way to distinguish these strategies, because a plant with long rhizomes can behave like a Guerrilla in hostile territory but so can a plant with short rhizomes that produces few offspring ramets, so its ramets are surrounded by plants of other species. Conversely, a plant with long rhizomes can easily form a monospecific stand that meets plants of another species at a sharp demarcation line.

The question also arises, what are long rhizomes? Reed (*Phragmites communis*) has a rhizome increment of one meter per year, but the plant can be two meters tall, so this great distance cannot help the offspring ramet escape shading by the parent plant; conversely, the arctic sedge (*Carex subspathacea*) is only 2–3 cm tall and yet has rhizomes up to 15 cm long, that offspring ramet cannot be shaded by the parent plant. This ratio between plant height and lateral spread per year could therefore be a better measure to distinguish Phalanx and Guerrilla strategies. We tried to see how this ratio changes according to environmental gradients, but we found no differences—larger plants generally have longer rhizomes, so the ratio between height and length remains approximately the same despite the given examples (Klimešová et al. 2011).

3.5.4 Integrators and Splitters

Integrators and splitters are extreme strategies of clonal plants in terms of how long the connection between ramets persists; we have already mentioned this trait in the chapter on clonal traits. This concept was created based on the ecology of several species of clonal plants by Inga Jónsdóttir and Maxime Watson in the late nineties (Jónsdóttir and Watson 1997) and was elaborated in detail in modelling studies by Beata Oborny (Oborny et al. 2000). It is based on the experience that in mild conditions with sufficient nutrients and moisture, plant clones disintegrate sooner than in

Fig. 3.34 Main strategies of clonal plants. Integrators are plants with long-term connection of ramets that are typical for dry habitats with nutrient scarcity. Splitters with short-term connection between ramets are typical for moist and nutrient-rich sites

stressful conditions of water and nutrient scarcity. This is attributed to the fact that large clones with interconnected ramets and an extensive system of rhizomes with roots are capable of harvesting enough resources from a large area of rooted soil to sustain the plant's life and share resources through the connected system (Fig. 3.34).

Despite individual examples of extensive clones growing in communities with low nutrient and water availability, it seems from correlative studies, based on data on lateral spread and the persistence of connections between ramets, that integrating clones are indeed more common in these conditions than splitting ones. Contrary to expectations, however, these are not extensive clones, but rather small clones with limited lateral spread (Klimešová et al. 2016b) or even non-clonal plants. Thus, nutrient harvesting over a large area of the community is probably not crucial for the strategy of integrated clones, and the reason for their prevalence in these conditions must be sought elsewhere. New Zealand ecologists Hay and Kelly point out that the cohesiveness of clones is evolutionarily older than their disintegration, and therefore we should rather look for reasons for splitting than for integration (Hay and Kelly 2008). The reason why splitting clones dominate in wet to moist, nutrient-rich places and integrated clones in dry and nutrient-poor places may reflect different needs to conserve (reuse) limiting resources. In wet and nutrient-rich conditions, there are also intense decomposing microbial processes, and plants would have to invest a lot of energy into protecting belowground organs to resist decomposition. They would have to reinforce them with lignin and fill them with secondary metabolites that inhibit microbial and fungal activity in conditions where it is easier to let old organs decompose and create new ones—after all, resources are abundant here.

Integrators and splitters are thus at the ends of the economic spectrum of rhizomes, just as deciduous and evergreen leaves are on opposite ends of the economic spectrum of leaves (see the chapter on functional traits). However, this economic spectrum of rhizomes has not been extensively studied yet.

3.6 The Plant as a Population of Shoots

3.6.1 What Is an Individual?

The architecture of a plant also affects its population dynamics and mainly causes the study of population dynamics in plants to be complex. We are often interested in how a particular species is doing and whether it has a chance to survive in a given location under current conditions. How to find out? The easiest way would be to count individuals in different age categories, and, based on knowledge of the age at which the species reproduces and how many offspring it produces, we would quickly find out how the species is doing. This approach is simple in human populations, where from the representation of individuals in age categories, we immediately know what would happen if current trends were maintained. However, in plants, we cannot use such an approach for all the reasons that come to mind: we often cannot recognize individuals, we are unable to determine age, we know little about the age of first reproduction or the number of offspring over a lifetime.

We have mentioned plant clonality many times, which causes the physically independent unit that is an individual within a population not to be identical to a genetically unique individual arising from a seed. When we count individuals of a species in a population, we usually find the number of shoots, which may or may not be genetically unique individuals (genets). If we dig up the belowground parts, we may find that several shoots are connected to one root; the plant is not clonal, but it has multiple shoots that appear to us above-ground as independent individuals. If such connected shoots have an abundance of adventitious roots belowground (or are themselves adventitious shoots, see the definition of clonality according to Groff and Kaplan), then we are dealing with a connected part of a clone, a clonal fragment (Fig. 3.35). Other clonal fragments belonging to the same genetic individual may be the nearest neighbors, but they can also be spread throughout the entire population. If we want to determine the ones belonging to a single genetic individual, we have no choice but to use molecular methods.

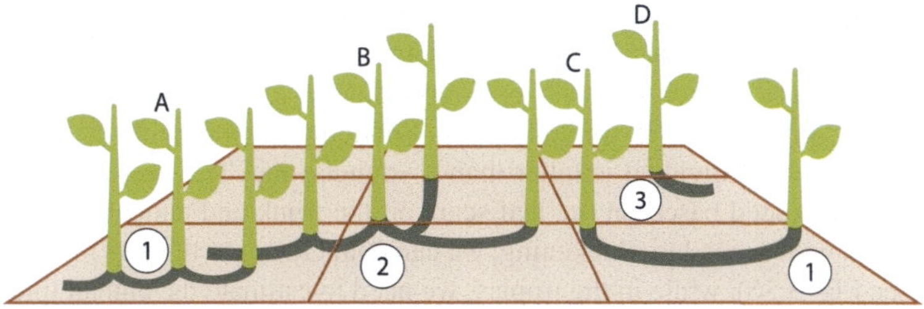

Fig. 3.35 Shoots that appear to be independent individuals above ground may be connected belowground by rhizomes and belong to different clonal fragments (**A**—**D**), which may or may not belong to a single genetic individual (genet) (1–3)

Fig. 3.36 How can the age of woody plants be measured? (**A**) a temperate zone tree Quercus as a dicotyledonous woody plant with secondary thickening has noticeable annual growth rings on the cross-section of the trunk, (**B**) palm (*Roystonea regia*, Mexico) as a monocotyledonous woody plant does not increase in thickness. (Photo: https://backyardnature.net/n/16/160124tq.jpg); (**C**) tropical dicotyledonous woody plant from Cameroon, thickness growth is measured by an automatic gauge. (Photo Jan Altman); (**D**) oak core prepared for measuring tree rings in the laboratory. (Photo Jan Altman)

3.6.2 Age and Size

In some plants, there is a simple relationship between the age and size of the plant. For example, in trees: the older the tree, the larger it is. More precisely, in trees that undergo secondary thickening, such as dicotyledonous trees, an older individual is wider than a younger one. In trees that do not undergo secondary thickening, such as palms, an older individual is taller than a younger one. The age of a palm can usually be calculated based on the leaf scars on the trunk and annual leaf production. In trees with secondary thickening, we can count the growth rings in temperate zones (see Fig. 3.36), while in the tropics, we need to estimate the annual thickness increment, for example, using an automatic meter attached to the trunk that regularly (e.g., daily) records the width increment (see Fig. 3.36). But how do we determine the age of trees that have multiple trunks that can replace each other, as is the case with poplars? For clonal trees, an estimate of the rate of clone spread could be used. Such estimates have been made for the aspen poplar in North America. However, it is necessary to determine, using molecular methods, which trees belong to a single clone. Here too, it probably holds that older clones are larger than younger clones, of course in the horizontal dimension, assuming that the clone continuously spreads evenly horizontally.

And how do we determine the age of an herbaceous plant? In herbs, the same approach as in trees can sometimes be used, that is, counting the growth rings on the oldest part of the body. In trees, growth rings are counted at breast height of the trunk, but in herbs, the transition between the primary stem and the primary root, including the hypocotyl, sometimes called the root crown, which is located belowground (see Fig. 3.37), is used for counting growth rings. The method can only be used in seasonal climates and only in herbs that have a root crown and growth rings are visible in the cross-section. Here too, it holds that the older the herb, the larger it is, with a wider root crown and main root, while the above-ground shoots do not enlarge after reaching the maturity of a perennial herb (roughly after the first flowering). However, in many herbs, the main root is often replaced by adventitious roots, thus the oldest part of the plant is not preserved, and the plant is clonal (see Section 3.3.1). The age of such a plant cannot be determined, even if the belowground stems have clear growth rings or morphological features that allow age counting. In this case, the size of the plant, measured as the thickness or length of the rhizome, does not reflect the age of the plant but rather how well it is doing in the community. If we count the growth rings (or morphological features reflecting annual increments) on the oldest part of the rhizome, we learn how old this part of the clone is, but the overall age of the plant originating from the seed (genet) must be determined differently (see Fig. 3.37). Again, we need to know the size of the entire genet, its lateral spread rate, and then we can calculate the approximate age. It is important to realize that the described methods are destructive and cannot always be used.

3.6.3 Small and Large Life Cycle of Perennial Plants

So far, we have dealt with the age of the entire plant originating from a seed, which is relatively easy to determine in a non-clonal plant but difficult in a clonal plant. We were interested in the length of the large life cycle, the time from seed germination to plant death. The large life cycle can be imagined, for example, in a tree as the germination of an acorn, from which a seedling grows, which continues to grow after winter. The first shoot thickens and branches, and thus the tree changes its size over the years, its trunk thickens, and branches and twigs (shoots) increase in number but not in size until, perhaps after 20 years from germination, it begins to bloom and bear fruit. Then it continues to grow and bear fruit until it dies sometime in 200–300 years. An herb starts its life similarly but bears fruit much earlier, perhaps as early as the second year of life. Each year it forms new shoots, which are always the same size, but there may be more or fewer of them until it dies in 20 or 30 years. What we have described is the large life cycle, the life cycle of an individual that originated from a seed (the life cycle of a genet, Fig. 3.38).

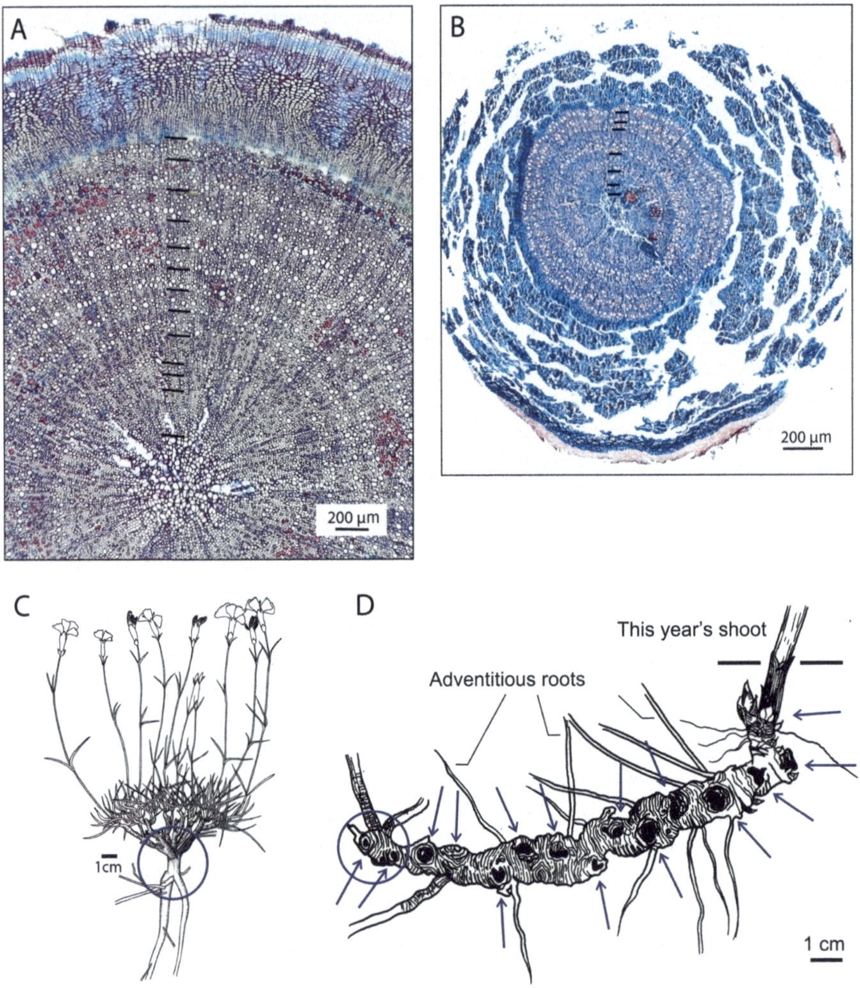

Fig. 3.37 Age of herbs according to growth rings (**A**, **B**) and rhizome increments (**D**). (**A**) Annual wood increments in the root crown of a 13-year-old *Lychnis viscaria* from a meadow near the village of Naramec (near Třebíč, photo Veronika Jandová), (**B**) Annual wood increments in the root crown of an 8-year-old *Draba oreades* from an altitude of 5700 m a.s.l. in the Himalayas. (Photo Veronika Jandová), (**C**) *Dianthus sylvestris* as an example of a non-clonal plant with the oldest part, the root crown, marked, (**D**) *Lunaria rediviva* as an example of a clonal plant with the oldest part of the rhizome marked. (According to Klimešová 2018). Because the rhizome forms regular annual increments, we can calculate that it survived 15 years (arrows), however, the age of the entire plant (genet) cannot be determined because the rhizome decays from its oldest part

Besides the life cycle of the entire genet, we can also observe the life cycle of a single shoot (the small life cycle) from the moment it grows from a bud to flowering and death. This life cycle is called cyclicity (see also the architecture of herbs). Cyclicity can be described in all plants, in trees, herbs, and shrubs, and their potentially flowering shoots. However, cyclicity is most important for the life strategy of

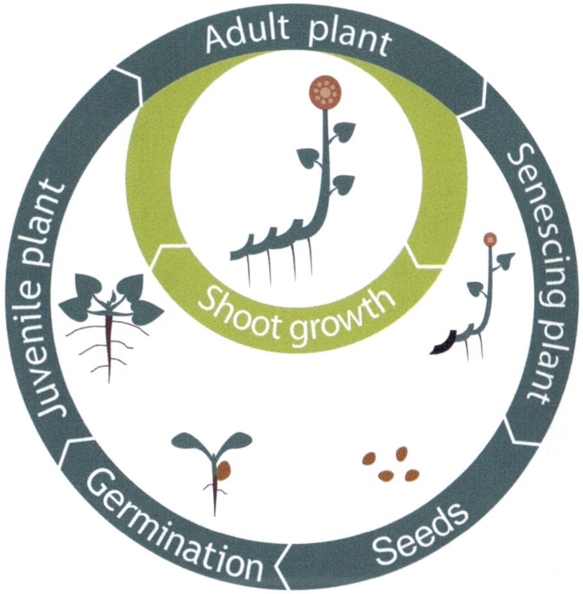

Fig. 3.38 Large (dark green) and small (light green) life cycles using the example of a clonal herb. The large life cycle describes the development of the plant from seed germination, through flowering, to aging and death; the small life cycle describes the development of the shoot from bud opening, through flowering to fruiting and death

a plant where the shoot is potentially an independent plant—that is, in clonal herbs and herbs with a single shoot, such as annuals and biennials. In non-clonal perennial herbs and woody plants, where shoots are part of larger functional units, the age of the entire plant is more important for spreading along environmental gradients than cyclicity. Both short-lived plants and plants with monocyclic shoots prefer environments with abundant resources compared to long-lived plants and plants with polycyclic shoots (Klimešová et al. 2016a).

While entire genets can be classified as monocarpic—plants that reproduce once in their lifetime, or polycarpic—plants that reproduce repeatedly, the shoots forming the plant (those potentially flowering) are always monocarpic. Shoots that flower in the first year of their life are called monocyclic, shoots flowering in the second year are called dicyclic, and shoots flowering even later are called polycyclic. An annual plant is always monocarpic and simultaneously has a monocyclic shoot or shoots (the small life cycle here equals the large life cycle). A biennial plant is monocarpic and has a dicyclic, or polycyclic if not strictly biennial, shoot or shoots (again, the small life cycle equals the large life cycle). A perennial herb is polycarpic, blooms repeatedly during its life, and its shoots can be monocyclic, dicyclic, or polycyclic. There are monocarpic perennial plants (some species of bamboo or palms) and they usually have polycyclic shoots. Plants with monopodial branching, which do not have special flowering shoots, but instead individual flowers (or inflorescences) in the axils of leaves of their potentially infinitely vegetatively growing shoot, do not fit into this scheme where we could speak of polycarpic shoots (for example, in white clover, *Trifolium repens*).

How the small life cycle, that is, the cyclicity of shoots, affects the growth of a clonal plant is explained in Fig. 3.39. We see that in an individual with monocyclic shoots, we always observe only offpsring shoots in the population because the parent

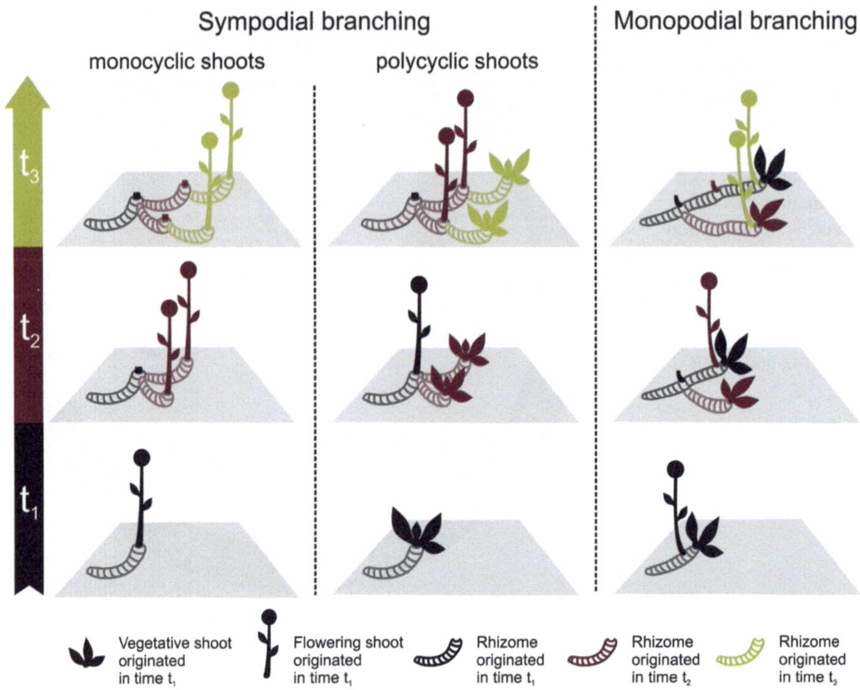

Fig. 3.39 Ontogeny of the shoot in clonal plants with different cyclicity. t1, t2, t3—three consecutive years. The left column shows a sympodially branching plant with monocyclic shoots, all of which can flower in a given year. The middle column shows a sympodially branching plant with dicyclic shoots, which flower in the second year of life, therefore in the population, we encounter flowering (shoots from the previous year) and non-flowering shoots (shoots from this year). The right column shows a plant that branches monopodially and has monocyclic flowering shoots, therefore in the population we encounter flowering and non-flowering shoots, but non-flowering shoots always remain non-flowering. (According to Janovský et al. 2017)

shoots are already dead and the generations of shoots do not overlap. This plant, for example, can share resources between parent and offspring ramets only through their storage in the rhizome. In contrast, in a plant with polycyclic shoots, the generations of ramets overlap and shoots of different ontogenetic stages can share resources directly; for example, a flowering shoot, which itself has few leaves, can be supported by its non-flowering offspring shoot. In a plant with monopodial branching, which has two types of shoots, one vegetative and one generative, such a division of labor can also be expected. However, the effect of shoot cyclicity on resource sharing or, for example, on the population growth of a clonal plant has not yet been studied, so the previous conclusions remain a mere hypothesis awaiting testing.

3.6.4 Population Dynamics

Now when we know what problems arise from understanding an individual and its age in plants, we must choose a simple and practical solution to describe the life cycle of plants and gain insight into population dynamics. Firstly, instead of age, we will use the ontogenetic stage and size of the plant, then determine whether the plant is clonal or non-clonal, practically identify individuals of all size categories, and monitor them every year for several years. Additionally, we will determine seed production for marked individuals and seed survival in the soil. We will record new individuals in the population and try to find out whether they originated from seeds or clonally, and also record the mortality of marked individuals. For clonal plants, we will determine (using molecular methods) which of the defined individuals belong to one genetic individual. Observations from year to year can be plotted in a table or diagram, where we obtain probability of transitions between size categories, natality from seeds and clonal growth, and mortality. From the information we gather about the given population, we can determine which processes are most important for the growth of our species' population, such as the survival of established individuals, regeneration from seeds, or clonal multiplication.

Although this method is quite labor-intensive, scientists have already gathered data for many plant species, and from comparing growth forms, we know that for short-lived non-clonal plants, the sensitive life stage is seed production and seedling survival, whereas for perennial herbs, it is persistence, and for trees, it is growth (Salguero-Gómez et al. 2016). When this key phase of the life cycle is threatened, for example, there is not enough competition-free space suitable for seedling establishment in short-lived herbs, the population begins to decrease. How does clonality fit into this scheme? Collecting the necessary data for clonal plants is not easy at all, and therefore they are recorded relatively rarely, especially species with extensive clonal spread (Janovský et al. 2017). This is due to problems in tracking marked individuals, for example, you do not know who is the parent of a newly emerged ramet when it suddenly appears half a meter from any potential parent ramets (molecular methods are not commonly used in these studies). Despite the methodological problems described, there are data on population dynamics even for clonal herbs, and the significance of clonality could thus be analyzed (Janovský and Herben 2020). According to this analysis, clonality appears more as a mechanism allowing population growth in favorable conditions rather than as a mechanism for surviving unfavorable conditions (mainly community disturbance). Clonality thus does not provide a universal advantage for population growth under all circumstances but allows plants to switch between regeneration from seeds in disturbed environments and clonal multiplication in undisturbed environments, utilizing the best of both regeneration methods.

References

Adler I, Barabe D, Jean RV (1997) A history of the study of Phyllotaxis. Ann Bot 80:231–244

Aeschimann D, Bocquet G (1980) Allorhizie et homorhizie, une reconsidération des définitions et de la terminologie. Candollea 35:19–35

Arber A (1920) Water plants. A study of aquatic Angiosperms. Cambridge University Press
Bartušková A, Klimešová J (2010) Reiteration in the short lived root-sprouting herb Rorippa palustris: does the origin of buds matter? Botany 88:630–638
Bartušková A, Lubbe CF, Qian J, Herben T, Klimešová J (2022) The effect of moisture, nutrients, and disturbance on storage organ size and persistence in temperate herbs. Funct Ecol 36:314–325
Bellingham PJ, Sparrow AD (2000) Resprouting as a life history strategy in woody plant communities. Oikos 89:409–416
Benson E, Hartnett DC (2006) The role of seed and vegetative reproduction in plant recruitment and demography in tallgrass prairie. Plant Ecol 187:163–177
Birch CPD, Hutchings MJ (1992) Analysis of ramet development in the stoloniferous herb Glechoma hederacea using a Plastochron Index. Oikos 63:387–394
Charles-Dominique T, Barczi JF, Le Roux E, Chamaillé-Jammes S (2017) The architectural design of trees protects them against large herbivores. Funct Ecol 31:1710–1717
Chelli S, Klimešová J, Tsakalos JT, Puglielli G (2024) Unraveling the clonal trait space: beyond aboveground and fine-root traits. J Ecol 212:730–740
Chomicki G, Coiro M, Renner SS (2017) Evolution and ecology of plant architecture: integrating insights from the fossil record, extant morphology, developmental genetics and phylogenies. Ann Bot 120:855–891
Corner EJH (1949) The durian theory or the origin of the modern tree. Ann Bot 13:367–414
Dalgleish HJ, Hartnett DC (2006) Below-ground bud banks increase along a precipitation gradient of the north American Great Plains: a test of the meristem limitation hypothesis. New Phytol 171:81–89
Diggle PK (1997) Extreme preformation in alpine Polygonum viviparum: an architectural and developmental analysis. Am J Bot 84:154–169
Ferris KG (2019) Endless forms most functional: uncovering the role of natural selection in the evolution of leaf shape. Am J Bot 106:1532–1535
Fidelis A, Appezzato-da-Glória B, Pillar VD, Pfadenhauer J (2014) Does disturbance affect bud bank size and belowground structures diversity in Brazilian subtropical grasslands? Flora 209:110–116
Groff PA, Kaplan DR (1988) The relation of root systems to shoot systems in vascular plants. Bot Rev 67:387–422
Hallé F, Oldeman RAA, Tomlinson PB (1978) Tropical trees and forests: an architectural analysis. Springer Verlag, Berlin
Harper JL (1977) Population biology of plants. Academic, London
Hay MJM, Kelly CK (2008) Have clonal plant biologists got it wrong? The case for changing the emphasis to disintegration. Evol Ecol 22:461–465
Herben T, Klimešová J, Chytrý M (2018) Effects of disturbance frequency and severity on plant traits: an assessment across a temperate flora. Funct Ecol 32:799–808
Huhta A-P, Hellström K, Rautio P, Tuomi J (2003) Grazing tolerance of Gentianella amarella and other monocarpic herbs: why is tolerance highest at low damage levels? Plant Ecol 166:49–61
Irmisch T (1850) Zur Morphologie der monokotylischen Knollen- und Zwibelgewachse. G. Reimer Verlag, Berlin
Janovský Z, Herben T (2020) Reaching similar goals by different means–differences in life-history strategies of clonal and non-clonal plants. Perspect Plant Ecol Evol Syst 44:125534
Janovský Z, Herben T, Klimešová J (2017) Accounting for clonality in comparative plant demography – growth or reproduction? Folia Geobot 52:433–442
Jeník J (1973) Root system of tropical trees 8. Stilt-roots and allied adaptations. Preslia 45:250–264
Jeník J (1978) Roots and root systems in tropical trees. In: Tomlinson PB, Zimmerman M (eds) Tropical trees as living systems. Cambridge University Press, pp 323–349
Jeník J (1994) Clonal growth in woody plants—a review. Folia Geobotanica Phytotaxonomica 29:291–306
Jones CS (1999) An essay on juvenility, phase change, and heteroblasty in seed plants. Int J Plant Sci 160:S105–S111

Jónsdóttir IS, Watson M (1997) Extensive physiological integration: an adaptive trait in resource-poor environments? In: de Kroon H, van Groenendael J (eds) The ecology and evolution of clonal plants. Backhuys Publishers, Leiden, pp 109–136

Klimeš L (1999) Small-scale plant mobility in a species-rich grassland. J Veg Sci 10:209–2018

Klimeš L, Klimešová J (1999) Root sprouting in Rumex acetosella under different nutrient levels. Plant Ecol 141:33–39

Klimešová J (2018) Temperate herbs: an architectural analysis. Academia, Praha

Klimešová J, Herben T (2015) Clonal and bud bank traits: patterns across temperate plant communities. J Veg Sci 26:243–253

Klimešová J, Klimeš L (2003) Resprouting of herbs in disturbed habitats: is it adequately described by Bellingham-Sparrow's model? Oikos 103:225–229

Klimešová J, Klimeš L (2007) Bud banks and their role in vegetative regeneration–a literature review and proposal for simple classification and assessment. Perspect Plant Ecol Evol Syst 8:115–129

Klimešová J, Klimeš L (2008) Clonal growth diversity and bud banks in the Czech flora: an evaluation using the CLO-PLA3 database. Preslia 80:255–275

Klimešová J, Martínková J (2022) Chapter 11: Clonal growth, resprouting, and vegetative propagation of weeds. In: Upadhyaya MK, Clements DR, Shrestha A (eds) Persistence strategies of weeds. Wiley Blackwell, pp 200–218

Klimešová J, Sosnová M, Martínková J (2007) Life-history variation in the short-lived herb Rorippa palustris: effect of germination date and injury timing. Plant Ecol 189:237–246

Klimešová J, Kociánová A, Martínková J (2008) Weeds that can do both tricks: vegetative versus generative regeneration of the short-lived root-sprouting herbs Rorippa palustris and Barbarea vulgaris. Weed Res 48:131–135

Klimešová J, Doležal J, Sammul M (2011) Evolutionary and organismic constraints on the relationship between spacer length and environmental conditions in clonal plants. Oikos 120:1110–1120

Klimešová J, Nobis M, Herben T (2016a) Links between shoot and plant longevity and plant economics spectrum: environmental and demographic implications. Perspect Plant Ecol Evol Syst 22:55–62

Klimešová J, Tackenberg O, Herben T (2016b) Herbs are different: clonal and bud bank traits can matter more than leaf–height–seed traits. New Phytol 210:13–17

Klimešová J, Herben T, Martínková J (2017) Disturbance is an important factor in the evolution and distribution of root-sprouting species. Evol Ecol 31:387–399

Klimešová J, Martínková J, Pausas JG, de Moraes MG, Herben T et al (2019) Handbook of standardized protocols for collecting plant modularity traits. Perspect Ecol Evol Syst 40:125485

Krasilnikov PK (1970) Klassifikacyia kornevych system derevev i kustarnikov. Lesovedenie 3:35–44

Krumbiegel A (1998) Growth forms of annual vascular plants in Central Europe. Nord J Bot 18:563–575

Krumbiegel A (1999) Growth forms of biennial and pluriennial vascular plants in Central Europe. Nord J Bot 19:217–226

Kutschera L (1960) Wurzelatlas mitteleuropäischer Ackerkräuter und Kulturpflanzen. DLG Verlags gmbh, Frankfurt am Main

Kutschera L, Lichtenegger E (1982) Wurzelatlas mitteleuropäischer Grünlandpflanzen, vol Band 1. Verlag Gustav Fischer, Stuttgart, Monocotyledoneae

Kutschera L, Lichtenegger E (1992) Wurzelatlas mitteleuropäischer Grünlandpflanzen. Band 2 Pteridophyta und Dicotyledoneae. Verlag Gustav Fischer, Stuttgart

Lovett-Doust L (1981) Population dynamics and local specialization in a clonal perennial (Ranunculus repens): I. The dynamics of ramets in contrasting habitats. J Ecol 69:743–755

Lux A, Baláž M, Kummerová M, Soukup A, Votrubová O et al (2017) Illustrated guide to plant anatomy. Academia, Prague

Martínková J, Klimešová J, Mihulka S (2008) Compensation of seed production after severe injury in the short-lived herb Barbarea vulgaris. Basic Appl Ecol 9:44–54

Martínková J, Klimeš A, Klimešová J (2018) No evidence for nutrient foraging in root-sprouting clonal plants. Basic Appl Ecol 28:27–36
Mogie M, Hutchings MJ (1990) Phylogeny, ontogeny and clonal growth in vascular plants. In: van Groenendael J, de Kroon H (eds) Clonal growth in plants: regulation and function. SPB Publishing, The Hague, pp 3–22
Mohan Ram HY, Sehgal A (1997) In vitro studies on developmental morphology of Indian Podostemaceae. Aquat Bot 57:97–132
Niklas K (1992) Plant biomechanics. In: An engineering approach to plant form and function. The University of Chicago Press, Chicago
Niklas KJ, Hammond ST (2013) Biophysical effects on plant competition and coexistence. Funct Ecol 27:854–864
Nishii K, Huang B, Wang C, Möller M (2017) From shoot to leaf: step-wise shifts in meristem and KNOX1 activity correlate with the evolution of a unifoliate body plan in Gesneriaceae. Dev Genes Evol 227:41–60
Oborny B, Kun Á, Czárán T, Bokros S (2000) The effect of clonal integration on plant competition for mosaic habitat space. Ecology 81:3291–3304
Ott JP, Klimešová J, Hartnett DC (2019) The ecology and significance of below-ground bud banks in plants. Ann Bot 123:1099–1118
Ottaviani G, Lubbe FC, Lepš J, Lisner A, Martínková J et al (2021) Strong impact of management regimes on rhizome biomass across central European temperate grasslands. Ecol Appl 31:e02317
Pausas JG, Lamont BB, Paula S, Appezzato-da-Glória B, Fidelis A (2018) Unearthing below-ground bud banks in fire-prone ecosystems. New Phytol 217:1435–1448
Pyne SJ (2017) The Great Plains. A fire survey. The University of Arizona Press
Qian J, Wang Z, Klimešová J, Lü X, Kuang W et al (2017) Differences in below-ground bud bank density and composition along a climatic gradient in the temperate steppe of northern China. Ann Bot 120:755–764
Rauh W (1937a) Beiträge zur Morphologie und Biologie der Holzgewächse. Die Entwicklungsgeschichte und Verzweigungsverhältnisse arktisch-alpiner Spaliersträucher. Nova Acta Leopoldiana N F 5:289–348
Rauh W (1937b) Die Bildung von Hypokotyl- und Wurzelsprossen und ihre Bedeutung für die Wuchsformen der Pflanzen. Nova Acta Leopoldiana N. F. 4:395–553
Rauh W (1939) Über polsterförmigen Wuchs, ein Beitrag zur Kenntnis der Wuchsformen der höheren Pflanzen. Nova Acta Leopoldiana N. F. 7:267–508
Salguero-Gómez R, Jones OR, Jongejans E, Blomberg SP, Hodgson DJ, Mbeau-Ache C, Zuidema PA, de Kroon H, Buckley YM (2016) Fast-slow continuum and reproductive strategies structure plant life-history variation worldwide. PNAS 113:230–235
Schnablová R, Herben T, Klimešová J (2017) Shoot apical meristem and plant body organization: a cross-species comparative study. Ann Bot 120:833–843
Schnablová R, Huang L, Klimešová J, Šmarda P, Herben T (2020) Inflorescence preformation prior to winter: a surprisingly widespread strategy that drives phenology of temperate perennial herbs. New Phytol 229:620–630
Serebriakova TI (1977) On major architectural models of herbaceous perennials and patterns of their reconfiguration. Biulleten Moskovskovo Obshczestva Ispytatielej Prirody, Otdel Biologii 82(5):112–128. (in Russian)
Šmilauerová M, Šmilauer P (2007) What youngsters say about adults: seedling roots reflect clonal traits of adult plants. J Ecol 95:406–413
Song M, Dong M (2002) Clonal plants and plant species diversity in wetland ecosystems in China. J Veg Sci 13:237–244
Sosnová M, Herben T, Martínková J, Bartušková A, Klimešová J (2014) To resprout or not to resprout? Modeling population dynamics of a root-sprouting monocarpic plant under various disturbance regimes. Plant Ecol 215:1245–1254
Stuefer JF (1998) Two types of division of labour in clonal plants: benefits, costs and constraints. Perspect Plant Ecol Evol Syst 1:47–60

Stueffer JF, de Kroon H, During HJ (1996) Exploitation of environmental hetergeneity by spatial division of labor in a clonal plant. Funct Ecol 10:328–334

Vanderweide BL, Hartnett DC (2015) Belowground bud bank response to grazing under severe, short-term drought. Oecologia 178:795–806

Verwijst T, Tavaziva VJ, Lunkvist A (2018) Assessment of the compensation point of Cirsium arvense and effects of competition, root weight and burial depth on below-ground dry weight – leaf stage trajectories. Weed Res 58:292–303

Weaver JE (1958) Classification of root systems of forbs of grassland and a consideration of their significance. Ecology 39:394–401

Weiser M, Koubek T, Herben T (2016) Root foraging performance and life-history traits. Front Plant Sci 7:779

Whitman T, Aarssen LW (2010) The leaf size/number trade-off in herbaceous angiosperms. J Plant Ecol 3:49–58

Wise MJ, Abrahamson WG (2008) Applying the limiting resource model to plant tolerance of apical meristem damage. Am Nat 172:635–647

Wright IJ, Dong N, Maire V, Prentice IC, Westoby M et al (2017) Global climatic drivers of leaf size. Science 357:917–921

Xie XF, Song YB, Zhang YL, Pan X, Dong M (2014) Phylogenetic meta-analysis of the functional traits of clonal plants foraging in changing environments. PLoS One 9(9):e107114

Ye X-H, Yu F-H, Dong M (2006) A trade-off between guerrilla and phalanx growth forms in Leymus secalinus under different nutrient supplies. Ann Bot 98:187–191

Ye D, Hu Y, Song M, Pan X, Xie X et al (2014) Clonality-climate relationships along latitudinal gradient across China: adaptation of clonality to environments. PLoS One 9(4):e94009

Zhang H, Bonser SP, Chen S-C, Hitchcock T, Moles AT (2018) Is the proportion of clonal species higher at higher latitudes in Australia? Austral Ecol 43:69–75

Conclusion

We have discussed the plant body from the overall appearance to the microscopic growth apex. We know that some shapes have an ecological function, we assume it for others, and for some, we are uncertain. In this book, I wanted to show that the description of form was available before understanding its function and that understanding function is a complex process that leads from random observation to hypothesis testing. It is important to realize that understanding the function of the entire diversity of shapes that plant bodies have is not possible without knowledge of this diversity and the methods of its description.

Where should we, as ecologists, focus our attention to help develop the field we can call functional morphology? While the understanding of the function of basic above-ground organs is relatively advanced, it is time to focus on the roots and derived belowground organs that ensure regenerative and clonal function. Even if we do not know how to use the classification of architectural models, there are concepts of plant architecture description that can be used in ecology. These include Corner's rule, the bud bank, clonal traits, shoot cyclicity, or reiteration. I hope this book has inspired you to utilize these concepts.

J. Klimešová, *Plant Bodies*, https://doi.org/10.1007/978-3-031-92192-6

Boxes

Box 1: Theophrastus (371 to 287 BC)

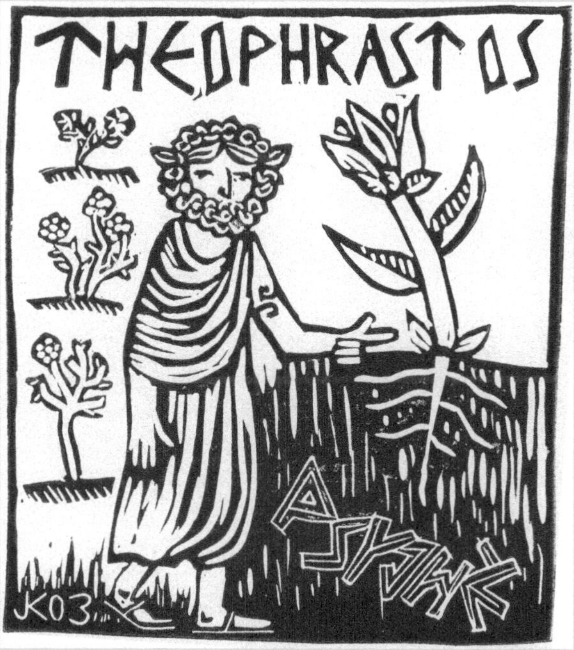

Theophrastus was an ancient Greek philosopher, a student of Plato and Aristotle, and is considered the founder of botany thanks to his works *Historia Plantarum* and *Causis Plantarum*, which have been almost entirely preserved. The Greek scholar Theodorus Gaza, who fled from his hometown Thessaloniki to Italy before the Turkish occupation, was responsible for their translation from Greek to Latin. Thanks to him, the books were reborn in 1483 and were a great inspiration for medieval scholars. Theophrastus dealt with all three

(continued)

J. Klimešová, *Plant Bodies*, https://doi.org/10.1007/978-3-031-92192-6

topics discussed in this book; he characterized growth forms and basic organs, and also noted how trees and herbs branch. For his detailed description of plant organs and the forms they can take, functional ecologists consider Theophrastus their model and the beginning of their field (Weiher et al. 1999). And given that we do not have older botanical works, functional ecology is thus one of the oldest botanical sciences.

Box 2: Alexander von Humboldt (1769–1859)

Humboldt was a German naturalist. He came from a wealthy family, was educated at home, and one of his teachers was the botanist Karl Ludwig Willdenow (later the director of the Berlin Botanical Garden, where some of Humboldt's collections ended up). Humboldt was interested in all natural sciences, and during his studies at the university, he met Georg Foster, who was

(continued)

a botanist on Cook's second voyage around the world. Together they traveled around Europe, and in England, they also visited the botanist Joseph Banks, who participated in Cook's first voyage to New Zealand and the eastern coast of Australia. Humboldt also befriended Goethe. They spent several months together in Jena, attended lectures, conducted experiments, and discussed science and art. Humboldt wanted to travel, but he was able to do so only after he could manage his property following his mother's death (his father died when he was 10 years old). The Spanish King Charles IV allowed him to travel to the Spanish colonies in South America. Humboldt then spent a substantial part of his fortune on his journey between 1799 and 1804.

From the journey, he published not only works on plants (including their forms (see image), and their distribution), meteorology, native languages, and culture, but also on how poorly the Spanish king treated the colonies. It was probably this critical work that closed further exploratory journeys for him. For example, he unsuccessfully tried to obtain permission from the English king to visit the Himalayas, where he wanted to verify his discoveries from the Andes. The only major journey he managed to undertake was in 1829 to Russia. The journey lasted 8 months and the expedition covered 15,500 kilometers. The expedition was sponsored by the Russian Tsar Nicholas I, with Humboldt's main task being to determine the possibilities of platinum mining. He was supposed to stick to the planned route and was forbidden to report on the social status of the subjects. The furthest point of the expedition was supposed to be Tobolsk, but Humboldt extended his journey to the Chinese border in the Altai. The rest of his life, Humboldt spent writing about his work. He was the greatest scientific celebrity of his time, and for a long time after his death. On the occasion of the 100th anniversary of his birth, celebrations were held all over the world. After the world wars, he was forgotten, as were many Germans, and today his significance is being rediscovered. In 2019, on the occasion of the 250th anniversary of his birth, several significant international scientific conferences were held, and his biography was published (Wulf 2016).

Box 3: Eugenius Warming (1841–1924)

Warming was a Danish naturalist. In 1859, he began studying at the University of Copenhagen, but interrupted his studies for 3 years to become the secretary of the Danish paleontologist Peter Wilhelm Lund, who lived and worked in the town of Lagoa Santa (near Belo Horizonte in the state of Minas Gerais) in Brazil. In his free time, he went on hiking expeditions in the surrounding area, during which he collected 2600 plant species, 370 of which turned out to be new to science. Upon his return, he sent herbarium specimens

(continued)

to more than 50 experts on individual taxonomic groups, and after 30 years, in 1892, he published the book Lagoa Santa (Warming 1892). It became the most comprehensive description of regional flora and vegetation in tropical areas for a long time, and was translated into Portuguese at the beginning of the twentieth century. In it, Warming discusses geomorphology, soil, climate, vegetation types, growth forms, phenology, weeds, succession, and the effects of drought and fires. His experiences in Brazil influenced his understanding of the various botanical fields he pursued: morphology, taxonomy, growth forms, plant geography, and especially ecology. The picture refers to this period and also recalls P. W. Lund's discovery, which he made in Lagoa Santa in 1843, and which reportedly pleased Darwin: in a karst cave, he found not only skeletal remains of various megafauna species but also human bones. Lund could never reconcile himself with the idea that humans lived simultaneously with extinct animals.

After returning from Brazil, Warming studied for a year in Munich and Bonn. He taught at the University of Copenhagen and in Stockholm. In 1885, he became a professor of botany at the University of Copenhagen, and director of the Copenhagen Botanical Garden. In 1895, he wrote the very first textbook on ecology, based on his course on plant geography. It was published in English only in 1909 under the title *Oecology of Plants—an introduction to the study of plant-communities.* You may not have heard of it, as it was written in Danish, and the German version by the ecologist Schimpler, who extensively drew from this textbook and other works by Warming, became much more well-known. He even used Warming's illustrations but did not acknowledge Warming's authorship or even cite him.

Box 4: Christen C. Raunkiær (1860–1938)

Raunkiær was a Danish botanist, and Warming's successor as a professor of botany at the University of Copenhagen. He was born as Christen Christiansen on the Raunkiær farm, using the farm's name as his surname when enrolling at the university. Before accepting the position of assistant at the Copenhagen Botanical Garden in 1893, he traveled around Denmark and the Netherlands, and later with his family around the Mediterranean and the Caribbean (Denmark had colonies there). His wife Ingeborg, who was a writer and painter, illustrated his botanical publications and published daring books on relationships between men and women for that time. They divorced in 1915, the same year their only son, a traveler in the Middle East, died of tuberculosis at just 25 years old.

Raunkiær was notable for his deep knowledge of flora and vegetation, and his belief that everything in nature that can be counted should be subjected to mathematical analysis. Besides the null model, which was to be compared with the spectrum of life forms, he dealt with species dominance in communities, and typifying communities according to leaf size. When he retired, he devoted himself to analyzing Danish poetry—and he approached this problem quantitatively as well, so thanks to him, Danes learned how many times dandelions were mentioned in Danish poetry.

The picture shows Raunkiær's life forms in winter when there is snow. Although this classification is used worldwide, the Danish winter was certainly a motivation for its creation.

Box 5: Rudolf Dostál (1885–1973)

Dostál was a Czech experimental botanist and a pioneer in the study of plant hormones. He was born in Eastern Bohemia and studied at the Faculty of Philosophy of the Czech University in Prague, where his teacher, Professor Bohumil Němec, introduced him to the experimental morphology of plants. He graduated in 1908 and then worked at various secondary schools. From 1915 to 1918, he taught at the grammar school in Třeboň, which is referenced by the image of the dam of the former fishpond, Hrádeček, where he collected plants for his experiments. During that time, Dostál completed his habilitation thesis: On the Formative Activity of Reserve Organs in Plants. After 1919, he worked at the University of Agriculture and the University of Veterinary Medicine in Brno. He continued his work even after retiring in 1955.

(continued)

Since his studies, he was engaged in the mutual regulation of growth between plant organs. His experimental work remained mostly unknown abroad, yet it influenced several researchers at the beginning of hormonal regulation research in plants. He published extensively, but mostly in Czech journals, and did not attend conferences. One of the pioneers of plant hormone research, American Kenneth V. Thimann, when traveling to a conference in the Soviet Union in 1961, decided to visit Rudolf Dostál in Brno and agreed with him to translate and publish Dostál's book, On the Integrity of the Plant, in America. He succeeded in 1965, and in his preface, he states that he considers the work not only inspirational for experimental developmental biology but also as a testimony to the influence of scientists in then-Czechoslovakia by Lysenkoism.[1] Dostál's support of Lysenkoism might explain the number of honors he received (Order of the Republic, Order of Labor, Klement Gottwald State Prize, and he was a member of the Czechoslovak Academy of Sciences and the Czechoslovak Agricultural Academy), but it is interesting that his student Jiří Šebánek, in his book about Dostál (Šebánek 2004), does not mention his inclination towards Lysenkoism at all. On the contrary, he emphasizes that Dostál was interested in nothing but his botanical work.

[1] Lysenkoism was a political campaign led by the Soviet biologist Trofim Lysenko (and supported by Stalin) against genetics (namely representatives of "formal genetics" or the so-called "Mendelism-Weismanism-Morganism") and science-based agriculture in the mid-twentieth century (Wikipedia)

Box 6: Wilhelm Troll (1897–1978)

Troll was a German plant morphologist. He was born in Munich, where he graduated in 1917 and left for the Western Front of World War I (1916–1918). After the war, he studied botany at the University of Munich, where his teacher was Karl von Goebel—a prominent morphologist, who combined morphology with physiology and plant anatomy. From 1928 to 1930, Troll participated in the natural science expedition "Sundaexpedition" to Sumatra, Java, the Moluccas, and Aru, where he, among other things, studied rooting in

(continued)

mangroves. In 1932, he became a professor of botany and head of the botanical garden in Halle, where he worked until 1945. In Halle, he rebuilt the botanical garden, trained many botanists, including Werner Rauh (see Box 18), and wrote his key work: Comparative Morphology of Plants. It is not the first work of its kind; already in the 1920s, Goebel published his Organography and Velenovský Comparative Morphology. However, it is definitely the last such extensive monograph, partly because it is exhaustive and also because after World War II, morphology began to address other issues.

His tenure in Halle coincided with the rise of fascism and World War II. During it, Troll managed to maintain funding for the botanical garden and teaching, and he was not drafted into the army because his superiors, and the publishing house publishing Comparative Morphology, interceded for him. It was probably his status as an important person that caused the American army to deport Troll (along with some other academics from the university) to the American zone after Germany's surrender. Although he was found to be politically blameless and not collaborating with the fascists, his return to Halle was not possible. Germany was divided into zones occupied by the victorious powers, and professors in the Soviet-occupied zone had to reapply for their positions, which was difficult for Troll from his place of involuntary exile in the western part of Germany (in the Palatinate). The University of Halle eventually found someone else, and Troll accepted an offer from the newly reestablishing university in Mainz, where he became a professor, head of the botany department, and botanical garden (all of which he had to build from scratch) (Von Nickel 1996). And perhaps, in the end, he was glad that it turned out this way…

Box 7: Agnes Arber (1879–1960)

Arber was an English botanist, morphologist, and philosopher. She was born in London as Agnes Robertson, studied in London and Cambridge, and for some time worked as an assistant in the laboratory of Ethel Sargant (one of the first women botanists in England). In 1909, she married geologist Arber (1870–1918), and in 1913, they had a daughter. From 1911 to 1927, she worked in the Balfour Laboratory at the women's colleges in Cambridge, and after its closure, she retreated to her private home laboratory until her death. She created it from a maid's room in the house where she lived with her

(continued)

daughter; the equipment was borrowed from the university. She led a modest life as a solitary researcher, being the sole author of most of her works (8 books and 93 scientific articles). At the beginning of World War II, she abandoned laboratory work and devoted herself to the history of botany and metaphysics. In 1946, she published a translation of her favorite work, Goethe's Metamorphosis of Plants (Box 8). After the war, she received professional recognition. She was the third woman, and the first botanist, ever elected as a member of the Royal Society in London, and the first woman to be awarded the Linnean Society's gold medal in London (Schmid 2001).

Agnes Arber primarily focused on the morphology of monocotyledonous plants, and illustrated her own books. At a time when scientific research moved from private settings to universities, Agnes Arber, writing in the dining room by the light of gas lamps, was the last of her kind (electricity was installed in the house only after her death). E.J.H. Corner (Box 13) wrote in her obituary that she was a lady of botany, although he himself resigned from several scientific societies in protest against their decision to admit women.

Box 8: Johann Wolfgang Goethe (1749–1832)

Goethe was a German writer, poet, and naturalist. His botanical essay "An Attempt to Explain the Metamorphosis of Plants" inspired the development of comparative morphology and plant architecture, and researchers in these fields translated his work into various languages (Arber into English, Serebriakova into Russian). Goethe admired Linnaeus's botanical system, and identifying plants according to this system was a social pastime at the court of the Duke of Saxe-Weimar, where he served as a lawyer and privy councilor. However, in his study of plants, he concluded that beyond identifying individual species, it was crucial to find one ideal archetype from which the entire diversity of forms in the plant kingdom is derived. He found the answer during his journey to Italy, where he was inspired by the dwarf fan palm (*Chamaerops humilis*) in the botanical garden in Padua (see picture). Humboldt (Box 2), who knew Goethe personally, spent some time with him in Weimar discussing science and the role of aesthetics in understanding nature, and often referred to his influence in his work.

Box 9: Carl von Linné (1707–1778)

Linnaeus was a prominent Swedish botanist, born in the southern Swedish province of Småland in the family of a pastor and farmer, who was himself an amateur botanist. When he entered university, he adopted the surname Linné—after the linden tree that grew in the courtyard of the family estate. It was expected that the eldest son, Carl, as was customary in the family, would study theology, but he chose botany and medicine. Thanks to the support of several individuals who provided him with free accommodation, access to lectures, and use of libraries, he graduated from Uppsala University.

In 1732, he received support from the Royal Academy of Uppsala and embarked on a journey through Lapland. The journey lasted 6 months,

(continued)

covered 2000 kilometers, and led around the Gulf of Bothnia from Uppsala to Turku, with several detours inland. The result of the journey was the description of 100 plant species new to science, and the book "Flora Lapponica", considered the first modern flora, which already contained the system of species description that Linnaeus later elaborated in his major works. On the journey through Lapland, he first saw a plant he became immensely fond of, which later received the name *Linnea borealis,* and which he used in his coat of arms after being ennobled. On his journey, Linné kept a diary, which was posthumously published in an English translation in 1811 (*Lachesis Lapponica: A Tour in Lapland*). He conducted several similar but smaller research trips across various regions of Sweden, mostly with students.

Between 1735 and 1738, Linné stayed abroad in Germany, the Netherlands, England, and France. In the Netherlands, he earned a doctorate (on a medical topic) and published his work Systema Naturae with the financial help of friends. In this and several subsequent works, he introduced new principles for classifying and naming plants. These principles have been significantly modified to this day but laid the foundation for modern botanical taxonomy. He abandoned naming plants based on their appearance and introduced two names: the genus, which indicates to which plants the species is most similar, and the specific epithet, which specifies exactly what species it is. He was offered to engage in botany in the Netherlands and travel to South Africa and America. Linné declined, stating it would be too hot for him there. After returning to Sweden in 1738, he never left the country again, but his 17 students (also called apostles) participated in various expeditions around the world. Seven of these botanists perished on their journeys (see image). Thanks to Linné's students, the method of description and classification system he devised spread worldwide. Moreover, this system proved effective in describing the vast number of species collected during exploratory journeys.

Box 10: Philip Grime (1935–2021)

Grime was a British ecologist who studied at the University of Sheffield and worked there in the Department of Comparative Ecology from 1961 to 2004 (he was its head from 1989). In the 1960s, he focused on describing vegetation, and he capitalized on the experience from this phase of his career in 1979 in the book Plant Strategies and Vegetation Processes, in which he introduced and described a system of plant strategies called Grime's Triangle (see image). In the subsequent period, the Department of Comparative Ecology, under his leadership, engaged in experimentation in the garden as part of the Integrated Screening Programme (1985–1997), which investigated the functional profiles of 43 common British herbs. The results were published in the book Comparative Plant Ecology (1988), which laid the foundation for functional ecology. Grime was also instrumental in the creation of one of the oldest experiments examining the impact of climate change on grassland communities (Buxton Climate Change Impacts Laboratory).

Box 11: CLO-PLA

The acronym CLO-PLA refers to a database containing data on clonal traits and bud bank characteristics for the flora of Central Europe. At a conference of researchers studying clonal plants in 1992 in the Krkonoše Mountains, we decided to quantify the types of clonal growth in our flora, to educate our colleagues who usually work on individual species about the diversity of clonal types. We wanted to warn them that they cannot generalize their conclusions from experiments on the creeping clonal plant *Glechoma hederacea* to all clonal species. By the next meeting of clonal plant researchers in Visegrád, Hungary, we prepared an overview of clonal organs for our flora, from which the CLO-PLA (clonal plants) database was created. We further refined this database, but we were still not satisfied with it, as it was not very functional. We had categorized plants according to clonal growth organs (rhizomes, stolons), and these categories were further divided into long rhizomes and short rhizomes, disintegrating clones, and non-disintegrating clones. However, the basis was still morphological categories, and the distinction of how far the clone spreads or how long the connection between ramets persists was secondary, which we found limiting and impractical for analyses. Therefore, we decided to build a database where the clonal growth organ would be just one of the clonal traits, but additional traits such as lateral spread of the clone, number of clonal offspring, or persistence of connection between ramets, among others, would be included. This resulted in a database that is now accessible on the internet and is part of other publicly accessible databases (TRY, PLADIAS).

Box 12: Caspar Wolff (1733–1794)

Wolff was a German physician and developmental biologist, born in Berlin, and in 1759 he completed his medical studies at the University of Halle with a dissertation “Theoria Generationis”. In this work, he deals with the embryonic development of plants and animals. Contrary to the prevailing opinion of the time, that the organism is already fully preformed in the seed (the theory of preformation), he supported the theory of epigenesis proposed by Aristotle. He demonstrated that the individual organs of animals are formed from undifferentiated cells, and in plants, he identified the growth apex as the source of these undifferentiated cells for organ formation. His views were not favorably received. During the Seven Years’ War, he served as a military doctor in the Prussian army and then, initially unsuccessfully, sought an academic position. Eventually, in 1767, he became the head of anatomy at the Academy of Sciences in St. Petersburg, where he lived until his death.

Box 13: Edred John Henry Corner (1906–1996)

Corner was a British botanist and mycologist. He studied at the University of Cambridge and published during his studies. Believing that a doctorate would not advance his knowledge, and that his stuttering made him unsuitable for a teaching career, he applied for a position as a mycologist at the botanical garden in Singapore. When he obtained the position, under the leadership of R. E. Holttum, he joined the colonial service and began his research on tropical rainforests in Singapore in 1929. He soon realized that he had to focus not only on fungi but also on plants, spending much time in the field and in the herbarium.

Corner wrote a book on common trees of Malaysia (published in 1940 under the title Wayside Trees of Malaya), which became an important identification tool for botanists throughout the Pacific. During his fieldwork, he was inspired by trained monkeys harvesting coconuts, and trained several to collect botanical samples in the treetops. In 1939, he married an American woman 13 years his junior after a brief acquaintance. Although the British in Singapore believed that Great Britain would protect them from a Japanese invasion during WW2, Singapore surrendered to the Japanese army at the beginning of 1942. Corner's wife and young son fled through Australia and America to England, and he remained in occupied Singapore. Unlike most Englishmen, he did not end up in a concentration camp, but with the governor's commission to protect the herbarium, museum, and botanical garden, he joined the occupiers' service. Together with several colleagues, this role was confirmed, and until the end of the war, he was part of a team that included Japanese scientists, tasked with preserving and further developing the

(continued)

collections during wartime. Although Corner was officially a prisoner, he was allowed to work in the herbaria, alongside Japanese scientists, mainly the botanist Koriba, and save collections, as well as private libraries or statues, from destruction by either side's forces.

After the war, Corner was repatriated to England, and because he was suspected of collaborating with the enemy, he was dismissed from the colonial service. Although he was rehabilitated 2 years later, he did not return to Singapore, and began working on a UNESCO project focused on researching the Amazon rainforest. After 2 years, during which he moved his family to Brazil and traveled through Latin American countries, he returned to England. From 1959, he was a lecturer, and from 1965, a professor at the University of Cambridge (Corner 2013). There, he wrote his evolutionary theory "The Durian Theory," which was soon rejected, and only Corner's rule, derived from the theory by Hallé and Oldeman and published in the first edition of the book on architectural models of trees, survived to this day. Interestingly, the three prominent botanists who spent the war together in occupied Singapore: Corner, Koriba, and Holttum, have their architectural types in Hallé's book (see picture).

Box 14: Leonardo of Pisa, Known as Fibonacci (Circa 1180–1250)

Leonardo was a mathematician, born in Pisa (which was then a city-state) sometime in the 1170s or 1180s. His father Guglielmo was nicknamed Bonaccio (the good-natured), and Leonardo was named Fibonacci after him (from filius Bonacci—son of Bonaccio). Leonardo's father was a high-ranking state official and head of a trading post in Béjaïa (a port located in present-day Algeria), and young Leonardo often traveled to Africa with him. It was probably on one of these trips that he became acquainted with the Arabic numeral system, and in 1202 he wrote the book Liber Abaci (The Book of Calculation), in which he introduced the Arabic numeral system to Europe. He provided very practical examples of its use, especially for merchants. At the same time, he introduced the concept of zero and the sequence of numbers known as the Fibonacci sequence into European mathematics. This sequence was created as a description of the growth of a rabbit population under ideal conditions (see picture).

Box 15: Charles Bennet (1720–1793)

Bennet was a Genevan lawyer, but he was also involved in natural sciences and philosophy. He discovered, for example, parthenogenesis in aphids and the exchange of air in plant leaves, and described the arrangement of leaves along the stem in plants, as well as visual hallucinations in people with poor vision. He lived with his wife near Geneva, and apparently never left this region during his lifetime. They had no children, and raised Mrs. Bennet's nephew as their own son. This nephew was Horace Bénédict de Saussure, who explored the Alps as a botanist, geographer, and geologist, measuring temperatures, the blueness of the sky, the boiling point of water, and humidity on various peaks and glaciers of the Alps, and in this, he was an inspiration for Alexander Humboldt (Box 2). Humboldt traveled with him through the Alps and took some of his improved meteorological instruments on his journey to South America.

Box 16: Thilo Irmisch (1816–1879)

Irmisch was a German naturalist, historian, and teacher. He studied at the University of Halle, and then worked, until his death, as a private tutor or high school professor in his hometown of Sonderhausen in Thuringia. He corresponded with about 50 botanists and also received botanical material for study from many of them. Several German universities invited him to apply for a professorship, but he did not respond to the invitations. Besides botany, he was also involved in history, archaeology, and local studies.

He collaborated with the morphologist and physiologist W. Hofmeister, and they planned a multi-volume work: Handbook of Physiological Botany (Handbuch der physiologischen Botanik), in which Irmisch was to write the volume on the Doctrine of Shoot Order (Die Lehre vor den Sprossfolge). Irmisch's contribution to this endeavor was found in his estate as a manuscript and was never published.

In 1850, Irmisch published a book on the growth of tuberous and bulbous plants, which is characterized by great meticulousness and includes, for example, an index and a list of literature, which were not yet common in botanical literature at that time. Irmisch published his works, among others, in the botanical journal Botanische Zeitung (see picture), which was published from 1840 to 1910. The journal also looked like a newspaper, was published weekly (!), and contained botanical observations, species lists, descriptions of plant growth, and short communications about botanists and plants. Longer articles often appeared in installments.

Box 17: Ivan Grigorievich Serebriakov (1914–1969)

Serebriakov was a Russian botanist. After completing his basic education and teacher training college, he started working at the age of 17 in his native village in central Russia (Ryazan region) as a teacher. Only after 3 years of teaching did he continue his studies at a pedagogical institute and Moscow University. During his studies, he spent two seasons in the Tian Shan, where he collected material for his diploma and doctoral thesis, which was on the topic of the biology and phytocenology of stands with *Picea schrenkiana*. He graduated in the field Geobotany (i.e., plant ecology) on the day the Soviet Union entered World War II. Although Serebriakov was exempted from military service due to heart failure, he volunteered for the front and was drafted. However, he was soon discharged as unfit and returned to the Moscow Botanical Garden and the university, where he taught for senior colleagues and defended his dissertation as early as 1943. In 1942, he married his

(continued)

colleague Tatyana Zaporina, who was 8 years younger (Box 20). After the war, he participated in an expedition to the Khibiny Mountains, mapping vegetation in the Polar Urals. It was in the Polar Urals that he learned that his employment at the Moscow Botanical Garden was terminated with immediate effect, and he and his family were evicted from their official apartment. The reason was the reckoning with representatives of "formal genetics" or the so-called "Mendelism-Weismanism-Morganism" and the facilitation of the development of Michurin science (Serebriakov had nothing to do with genetics) under Lysenkoism.[2] Since then, Serebriakov worked at the Pedagogical Institute in Moscow as the head of the Department of Botany.

Although Serebriakov studied geobotany, working in the botanical garden under the guidance of plant morphologist K.I. Mejer led him to plant morphology. He mainly focused on the ontogeny and seasonal development of herbs, inspired mainly by the works of German authors such as Irmisch, Goebel, and Troll, but also the Czech Velenovský. He had to renounce all of them in the introduction of his fundamental book "Morphology of Vegetative Organs of Higher Plants" (1952) to confirm that his views were in line with Michurin science. Troll, who was then a banned author in the Soviet Union, is called a proponent of the "aggressively idealistic direction." Although Serebriakov was a convinced communist, he did not succumb to Lysenkoism, and according to the testimony of close family members and colleagues, the introduction of his book appears as a mandatory offering so that the book could be published at all at that time (see picture) (Shafranova 2004).

An important milestone for him was participating in the International Botanical Congress in Paris in 1954, with an excursion to equatorial Africa. A personal turning point was a stroke that struck him in 1956 when he was 41 years old, and paralyzed the left side of his body. Nevertheless, he was able to partially regain movement in his arm and leg and returned to work, with great help from his wife. In 1962, his book "Ecological Morphology of Plants" was published. He taught, led many students, and his approach was widely used by his colleagues, but his health continued to deteriorate until he died in 1969. In Russia, Serebriakov was very influential, but he remained unknown beyond its borders.

[2] Lysenkoism was a political campaign led by the Soviet biologist Trofim Lysenko (and supported by Stalin) against genetics (namely representatives of "formal genetics" or the so-called "Mendelism-Weismanism-Morganism") and science-based agriculture in the mid-twentieth century (Wikipedia)

Box 18: Werner Rauh (1914–2000)

Rauh was a German botanist who completed his studies at the University of Halle under the guidance of Wilhelm Troll and stayed there for two more years, during which he managed to participate in expeditions to Tibet, Corsica, and the Swiss Alps. From 1939, he worked at the University of Heidelberg, where he habilitated in 1940 with a thesis on cushion growth forms. In this work, he utilized his own experiences from the Alps and Tibet, and plant material sent to him by affiliated botanists from South America and New Zealand.

During World War II, he was first with the infantry and then he was a meteorologist with the navy. From 1960, he was the head of the Institute of Plant Systematics and Geography, and the director of the botanical garden. He was essentially a plant hunter, mainly of tropical succulents, epiphytes, and high-altitude plants of the Andes. Almost every year, he participated in some expedition, working mainly in Madagascar and South America, but he also traveled extensively elsewhere in the tropics. On 66 trips of various lengths, he spent 114 months collecting plants for the botanical garden and herbaria (Schroder 2013)—today, such carefree collections of large numbers of plants and their transport to Europe are no longer possible. He discovered about 1200 taxa (species, subspecies, and varieties) new for science, and many of them were named in his honor. Although his work under Troll's guidance was morphological, he later focused mainly on systematics. Among his more than 300 publications (including over 20 books), we also find the booklet "Growth Forms of the High Andes," which I have in my library. In the 1990s, I requested

(continued)

it as part of the then-common system of requests for reprints—an article copy separated from the rest of the journal in which it was published. Scientists used to send these to each other by mail in those pre-internet days. Instead of the expected article, I received the entire book. I was so grateful to the author that I sent him a drawing of a winter aconite (*Eranthis hyemalis*) in return. At that time, I did not know that he himself was a skilled plant illustrator with hundreds of illustrations to his name.

Box 19: Francis Hallé (Born 1938)

Hallé is a French botanist, born into a botanical-artistic family, who studied at the Sorbonne and the University of Abidjan (Ivory Coast), where he worked from 1960 to 1968. He then stayed in Zaire, Congo, Indonesia, and visited many other countries, working at the University of Montpellier. He studied tree growth by drawing them, which led him to understand their architecture and describe architectural models of trees. In the 1980s, he began researching tree canopies using a raft—a balloon-supported platform in the canopies of the tropical rainforest. Now retired, he is involved in popularizing the protection of tropical rainforests, exhibiting his drawings, and publishing popular science books. I visited him in his apartment in an old building surrounded by trees in the center of Montpellier in January 2019 to thank him for inspiring my work on the architecture of herbs. We overcame initial distrust (on his part) and shyness (on my part) and soon delved into a conversation about plant morphology. Francis Hallé ran to his upstairs library for more and more books, and I admired his indoor tropical forest, the view from the balcony, and the paintings on the walls, painted by his grandfather, the notable French painter André Dauchez (see picture).

Box 20: Tatiana Ivanovna Serebriakova (1922–1986)

Serebriakova was a Russian botanist, born in Moscow as T. I. Zaporina, and from childhood attended a natural science club at the Moscow Botanical Garden led by A. V. Kozhevnikov. In 1939, she began studying at Lomonosov University, and in 1942 married I. G. Serebriakov. She closely collaborated with her husband I. G. Serebriakov (Box 17), and her plant drawings can be found in publications where she is not a co-author. After her husband suffered a stroke in 1956, she cared for him until his death in 1969, helping him publish and mentor students. Initially paralyzed on one side, his condition later improved somewhat, allowing him to lecture and go into the field. After her husband's death, she led the department and edited several collections on plant biomorphology. She admired Goethe and Agnes Arber, translated "The Metamorphosis of Plants" into Russian, and used a quote from Agnes Arber at the beginning of her 1971 book on grasses.

It is difficult to separate her personal contribution to the concept of biomorphology promoted by Serebriakov. Undoubtedly, she contributed to the description of life forms of grasses ("Morphogenesis of Shoots and Evolution of Life Forms of Grasses," 1971) and to defining the basic architectural models of herbs following Hallé and Oldeman in 1977. She also contributed to the synthesis of the biomorphological approach to plant ontogeny of the

(continued)

Serebriakov school, and the population approach of A. A. Uranov, which originally came together purely by administrative intervention, namely the merging of the botany departments of two pedagogical institutes (Zhukova 2015). This approach led to the description of the ontogeny of many hundreds of species from the Russian flora, and this activity continues to this day. Although determining population dynamics by the spectrum of life stages or evolutionary schemes of life forms in individual taxonomic groups (without the support of molecular methods) are now outdated, descriptions of ontogeny deserve more attention from population biologists.

Serebriakova‘s book on grasses was the first morphological work I ever read, and it allowed me to understand the growth of grasses, which I studied in my diploma and doctoral work (*Nardus stricta* and *Phalaris arundinacea*). When I started collecting data for the CLO-PLA clonal plant database, I initially primarily used the Russian works of the Serebriakov couple, their collaborators, and students, and only later did I learn German well enough to utilize plant descriptions from German researchers. This allowed me to study the plant’s form in its dynamic shape right from the beginning.

Box 21: Lore Kutschera (1917–2008)

Kutschera was an Austrian botanist, born in Villach as Eleonore Belani, and graduated from the University of Agricultural Sciences (University of Natural Resources and Life Sciences) in Vienna (1939). Until 1961, she worked in various positions in Lower Austria, usually involving agricultural consultancy. A decisive moment in her career was a study stay at the University of Nebraska, to which she was invited by the Association of Women Academics in 1954. From 1961 until her retirement in 1982, she was the head of the Department of Botany and Plant Sociology at the Federal Institute for Alpine Agriculture in the small town of Irdning in the Austrian state of Styria.

In 1942, she got married, but her husband did not return from the front and was declared missing after the war. In 1978, she became a docent. To facilitate the exchange of information among scientists dedicated to root systems, she founded the International Society of Root Research in 1982, where she was elected vice-president and organized several congresses. Her atlas of root systems of Central European plants is unparalleled in the world, and she, in collaboration with her colleague Erwin Lichtenegger, covered a substantial part of the herbaceous species of Central Europe (see picture). She also studied the influence of external conditions and management on the root systems of crops and weeds.

Box 22: Jan Jeník (1929–2022)

Jeník was a Czech ecologist. After studying forestry engineering, he joined the Department of Botany at Charles University as an assistant, and from 1957, as the head of the Geobotany department, where he habilitated in 1961 with a thesis on the vegetation of the High Sudetes. During the political relaxation of the 1960s, he sought to go abroad, which he succeeded in doing, and in 1965 he spent several months at the University of Kabul, and from 1966 to 1968 at the University of Legon (Ghana). The main result of his stay in Africa was the classification of the architecture of root systems of woody plants, which he based on the work of Russian botanist Krasilnikov, and enriched with specific studies of the root systems of several tropical trees. He returned to normalization after the fall of the Prague Spring and the tightening of the regime. He had to leave the university, initially employed at the Botanical Institute in Průhonice, and from 1975 in Třeboň.

(continued)

At the peak of his strength and work expansion, he was removed to a small town, where after pioneering research on vegetation in Czech mountains and the root systems of trees in Central Europe and mainly in the tropics, and after writing a textbook on tropical forests, he gradually reoriented towards popular science books for children, pictorial atlases, conservation and linguistic reflections, and organization of projects. He participated in leading a project in which the entire Třeboň workplace of the Botanical Institute worked on the Wet Meadows near Třeboň. Privately, with former students, the Bureš couple, he conducted research in the Velká Kotlina cirque in the Hrubý Jeseník, which was supposed to result in a monograph that, although written, was never published. Symbolic of the normalization period is Jeník's short article in Czech popular magazine Živa titled "The Whipped Form of Douglas Fir," in which he describes the growth of two trees on the grounds of the Botanical Institute in Třeboň: a fast-growing birch obstructs the growth of the Douglas fir with its crown, as it whips it with its branches and destroys buds and shoots, the Douglas fir is bent under the birch (see picture). In conclusion, he emphasizes that pioneer trees must be removed from the stand in time for the target tree to develop properly. I do not want to attribute symbolic meaning to the piece, which is indeed suggested, but the author may not have intended it. For me, the article is an expression of the despair of a person confined at home, suffering from a lack of objects to observe, while he would so much like to explore the world. When Jeník was 60 years old, the Velvet Revolution took place, and he was finally able to return to the university. He was rehabilitated, received the title of professor, and for another 5 years, he led the Geobotany department at the Department of Botany at Charles University in Prague.

Box 23: John Harper (1925–2009)

Harper was a British ecologist, studied at Oxford, and then worked there in agricultural research. After spending a year at the University of Davis (USA), he became the head of the School of Plant Biology at the University of Bangor (UK). He was the author or co-author of several very popular ecology textbooks used worldwide. I was most influenced by his book "Plant Population Biology" (1977), which opened up the world of plant population biology to me. In this book, the bud bank is briefly mentioned for the first time, while the author discusses the seed bank and its significance for population renewal in great detail.

References

Corner JK (2013) My father in his suitcase: in search of E.J.H. Corner the relentless botanist. Landmark Books

Schmid R (2001) Agnes Arber, nee Robertson (1879–1960): fragments of her life, including her place in biology and in women's studies. Ann Bot 88:1105–1128

Schroder CN (2013) Von den Mulde-Auen zur Cordillera Blanca—Zum 100. Geburtstag des Botanikers Werner Rauh (*16. Mai 1913 †7. April 2000). Die Bromelie 2013(2):63–69, 2013(3): 123–129

Šebánek J (2004) Harmonie v rostlinách. Academia, Praha

Shafranova LM (2004) Ivan Grigorievich Serebriakov, human and researcher. Prometej, Moskva (in Russian)

Von Nickel G (1996) Wilhelm Troll (1897–1978) Eine Biographie. Acta Hist Leopoldina 25:1–240

Warming E (1892) Lagoa Santa: Et Bidrag til den biologiske Plantegeografi med en Fortegnelse over Lagoa Santas Hvirveldyr. Kongelige Danske Videnskabernes Selskabs Skrifter – Naturvidenskabelig og Mathematisk Afdeling 6:153–488

Weiher E, van der Werf A, Thompson K, Roderick M, Garnier E, Eriksson O (1999) Challenging Theophrastus: a common core list of plant traits for functional ecology. J Veg Sci 10:609–620

Wulf A (2016) The invention of nature. The adventures of Alexander von Humboldt, the lost hero of science. Hodder & Stoughton

Zhukova LA (2015) Tatyana Ivanovna Serebriakova – the outstanding biomorphologist of the XX century. Samarskaya Luka: Problemy regionalnoj i globalnoj ekologii 24:213–228

J. Klimešová, *Plant Bodies*, https://doi.org/10.1007/978-3-031-92192-6

Index

J. Klimešová, *Plant Bodies*, https://doi.org/10.1007/978-3-031-92192-6

The manufacturer's authorised representative in the EU is Springer Nature Customer Service Centre GmbH, Europaplatz 3, 69115 Heidelberg, Germany. If you have any concerns regarding our products, please contact ProductSafety@springernature.com

Printed and bound by CPI Group (UK) Ltd, Croydon, CR0 4YY

29/07/2026

02180503-0009